QA
279
.L38
1993

Straight Talk on
Designing Experiments

Straight Talk on Designing Experiments

●●●●●●●●●●●●●●●●

An introductory design of experiments reference handbook

Robert G. Launsby
Daniel L. Weese

Launsby Consulting
Colorado Springs, CO

Cover design: Lee Brown

Text design: Diane Launsby
Ronda Churchill

We would like to thank the following people for their assistance in this project: Diane Launsby, Ronda Churchill, Lynda Coleman, Jayme Lahey, Lee Brown and Steve Dziuban, Ph.D. A special thanks to Brad Jones, Rich Shaw, Ron Boehly, and Susan Blake for providing case studies and other material for this text.

© Copyright 1993 LAUNSBY CONSULTING 1-800-788-4363

All rights reserved. No part of this book may be reproduced, stored in a retrieval system, or transcribed, in any form or by any means, electronic, mechanical, photocopying, recording, or otherwise, without the prior written permission of the publisher, Launsby Consulting, 2167 North Academy Blvd., Colorado Springs, Colorado 80909.

Printed in the United States of America

Library of Congress Catalog Card Number: 93-91423

ISBN 0-9636093-3-5

About the Authors

Robert G. Launsby, M.S. Engineering. Bob is President of Launsby Consulting. He has over twenty years of industrial experience in management, engineering, training, and applications of problem solving approaches. Bob rose to the rank of Consultant Engineer at Digital Equipment Corporation because of his innovative work in the application of statistical tools to engineering problems. Over the last fourteen years, he has led over five hundred applications of statistical methods in a wide variety of industries. Possessing strong presentation and facilitation skills, Bob has instructed over three thousand people in the use of statistical problem solving tools. Bob has taught seminars for Drexel University, Washington State University, University of Phoenix, University of Colorado, University of Nebraska, and the University of Northern Colorado. Bob is co-author of the text *Understanding Industrial Designed Experiments*.

Daniel L. Weese, M.S. Applied Mathematics, Harvard University. Dan is a Senior Consultant with Launsby Consulting. He served in the U.S. Air Force for twelve years as a Lead Scientific Analyst, a Contractor Cost/Schedule Analyst, and Assistant Professor at the U.S. Air Force Academy. Over the last five years, he has instructed over one thousand people in all levels of mathematics. During his tenure at the USAF Academy, Dan was named the Outstanding Military Educator for the Department of Mathematical Sciences.

Preface

We tell our clients that the only way they will survive as a world class competitor is to efficiently and effectively apply experimental design techniques. We are convinced these techniques will be best learned when applied as soon as possible after their introduction. However, we are all familiar with the black hole we fall into when we return to our desk after being gone a few days. Before we know it, hours turn into days and days into weeks, and we have not yet tried this useful new technique that might actually do us some good. Remembering what was said in a seminar a month or two earlier can be a challenge.

This introductory handbook is intended to fill in the gaps and give it to you "straight." Using the statapult, a wooden training device, we lead you through the basics of planning an experiment, selecting an orthogonal (or nearly orthogonal) array, conducting the experiment, analyzing data (and interpreting computer results), putting the process on target, and confirming results. We have used simple mathematics whenever possible and the almighty computer when it is not. Appendix B includes some matrix math to give those interested a better basis for interpreting computer results.

For those who have not had formal training on designing experiments, this text will get you started. A reality of effectively using these techniques is that everyone in your company must be involved. Too frequently pockets of people in an organization are doing a great job with these tools, ... but don't pass the word. We encourage you to pass this book around, along with your success stories. This is a book that everyone can understand. As a matter of fact, we believe it is what has been missing — straight talk on designing experiments.

BOB LAUNSBY   DAN WEESE

Table of Contents

I. Introduction - "The Need"   1

II. The Basics (Screening/Troubleshooting)   II-1

   Chapter 1   Planning   1-1

   Chapter 2   Selecting an Orthogonal Design
- 2.1 Design Matrix   2-1
- 2.2 Orthogonal Designs   2-3
  - 2.2.1 Fractional Factorials and Aliasing   2-5
  - 2.2.2 Resolution   2-7
  - 2.2.3 Tabled Taguchi Designs   2-7
  - 2.2.4 Other Designs   2-17
- 2.3 Nearly Orthogonal Designs — the D-Optimal   2-19
- 2.4 Orthogonal vs What???   2-23
- 2.5 Summary   2-25
  - 2.5.1 Two-Level Designs   2-26
  - 2.5.2 Three-Level Designs   2-27

   Chapter 3   Conducting Your Experiment
- 3.1 The Worksheet   3-1
  - 3.1.1 Randomization   3-3
  - 3.1.2 Repetition vs Replication   3-3
- 3.2 Common Errors   3-4
  - 3.2.1 Lack of Experimental Discipline   3-5
  - 3.2.2 Measurement Error/Too Much Variation   3-5
  - 3.2.3 Aliased Effect/Inadequate Model   3-5
  - 3.2.4 Something Changed   3-6
  - 3.2.5 Improper Experimental Region   3-6
- 3.3 Operator Instructions   3-6

Chapter 4	Analyzing Your Results		4-1
	4.1	Analysis of Means and Graphical Analysis	4-3
	4.2	Interpreting Computer Results	4-14
		4.2.1 Familiar Items	4-16
		4.2.2 Grouping Data	4-16
		4.2.3 Standard Error of the Estimate	4-20
		4.2.4 Degrees of Freedom	4-23
		4.2.5 Mean Square Between/Regression	4-23
		4.2.6 Distributions	4-27
		4.2.7 F Statistic	4-30
		4.2.8 Type I Error (α)	4-33
		4.2.9 P Value	4-34
		4.2.10 T-Statistic/Standard Error of the Coefficient	4-35
		4.2.11 Standard Coefficient	4-37
		4.2.12 Multiple R: .990	4-37
		4.2.13 Squared Multiple R: .979	4-37
		4.2.14 Adjusted Squared Multiple R: .970	4-40
		4.2.15 Tolerance	4-41
	4.3	Summary	4-41
Chapter 5	Putting Your Process on Target (with Minimum Variation)		5-1
	5.1	Apply Settings from Plots of Averages	5-2
	5.2	Set Prediction Equation Equal to Target and Find Settings	5-3
	5.3	Use Partial Derivatives to Locate Optima	5-4
	5.4	Contour Plots	5-5
Chapter 6	Confirmation		6-1
Case Study II-1			
Case Study II-2			

III. Robust Design

Chapter 7	Introduction to Robust Designs		
	7.1	What is a Robust Design?	7-1

7.2	Identifying Factors for a Robust Design	7-1
	7.2.1 Control Factors	7-2
	7.2.2 Noise Factors	7-3
7.3	When Should You Do Robust Design?	7-4

Chapter 8	**Applying the Box and Bubble to a Robust Design**	
8.1	Planning	8-1
8.2	Selecting an Orthogonal Array	8-3
8.3	Conducting the Experiment	8-7
8.4	Analyzing Your Results	8-7
	8.4.1 The Blended Approach	8-7
	8.4.2 The Taguchi Approach	8-10
8.5	Put Your Process on Target and Confirm	8-13
	8.5.1 The Blended Approach	8-13
	8.5.2 The Taguchi Approach	8-13

Chapter 9	**Products with Dynamic Characteristics**	9-1
9.1	Planning	9-3
9.2	Selecting an Orthogonal Array	9-3
	9.2.1 The Blended Approach	9-3
	9.2.2 The Taguchi Approach	9-3
9.3	Conducting Your Experiment	9-4
9.4	Analyzing the Results	9-4
	9.4.1 The Blended Approach	9-5
	9.4.2 The Taguchi Approach	9-6
9.5	Put the Process on Target and Confirm	9-10
	9.5.1 Blended Approach	9-10
	9.5.2 Taguchi Approach	9-10
9.6	An Example	9-10

Case Study III-1: Robust Design

IV. Modeling Designs

Chapter 10	**Central Composite Designs**	
10.1	Planning	10-1

    10.2   Selecting and Orthogonal Array   10-4
    10.3   Conducting the Experiment and Analyzing Your Results   10-9
    10.4   Conclusion   10-15
Case Study IV-1: Central Composite Design

Appendix A: Decoding Prediction Equations

Appendix B: Matrix Solution to Least Squares Method

Index

I. Introduction - "The Need"

Introduction

Richard Nixon was President. America was relieved the Viet Nam war had ended. Detroit was king in the auto industry. IBM and Univac were big names in the computer business. The Doobie Brothers sang "China Grove" and "Long Train Runnin'". In German Universities, U.S. companies were studied as the ideal organizations. Most American companies had to compete only with other American companies. Inefficiencies in product design/development and manufacturing were not a big deal. Costs could always be passed on to the customers. It was 1970; complacency reigned.

Ronald Reagan was President. There was the Laffer Curve. Quality Circles were hot. Deming was finally being listened to by U.S. Companies. Phil Crosby told us that "Quality is Free." Asian companies were making dramatic inroads into the automotive, steel, and consumer electronics markets. Higher quality and lower costs were attracting U.S. consumers to foreign produced products. U.S. companies began moving jobs off shore to take advantage of lower labor costs and more quality conscious workers. It was early 1980. Cost and quality replaced complacency as the driving force.

In the 1990's, high quality and low cost continued as key issues. A third issue, however, became important. Time-to-market became critical. As shown in the following table [1], companies with quick response time across their organizations outperform their slower competitors in terms of growth and profitability. For example, Wal-Mart replenishes its stock in a typical store every two days. Competitors replenish their stock every two weeks. This allows Wal-Mart to provide the same service levels with one-

Company	Business	Response Difference	Growth Advantage vs. Average Competitor	Profit Advantage vs. Average Competitor
Wal-Mart	Discount stores	80%	3×	2×
Atlas Door	Industrial doors	66%	3×	5×
Ralph Wilson Plastics	Decorative laminates	75%	3×	4×
Thomasville	Furniture	70%	4×	2×
Citicorp	Mortgage	85%	3×	N/A

fourth the inventory investment and offer its customers many times the choice of stock for the same investment in inventory as their competition. In another text [2], the impact of being a fast-cycle competitor vs. a slow-cycle competitor relative to technological leadership is discussed. As the graphics on the following page display, the fast-cycle competitor continuously widens the gap over the slow-cycle competitor. This technological gap translates into profits and long term survivability in the marketplace. For example, during the mid-80's, it was not uncommon to take five years to design a new computer disk drive. By the early 1990's, leading companies were delivering leveraged products in as little as twelve to sixteen months. Companies with longer development and design cycles were no longer competitive. Quality, cost, and time-to-market became key to survival in the global economy of the 1990's.

As consultants, we have a unique perspective on what is happening across a wide spectrum of U.S. industry. Everywhere we go we hear the following from engineering managers and engineers.

"Compared to only five years ago, we must develop new products in half the time (or less), the job must be done with fewer people, and the product must be right with the first shipment."

To meet this challenge, traditional approaches will no longer work. Engineers and scientists need new tools to help them make better, faster decisions about product concepts, designs, and processes. The premier tool in allowing technical groups to make this transformation is experimental design. When properly applied, our clients see a 50% (or more) improvement in their efficiency and effectiveness.

Changes must take place, however, in how experimental design techniques are taught and how they are applied. Since the early 1980's, great interest has been given to this approach. Many engineers and scientists have been trained. Unfortunately, not all who attend training use the techniques. We believe there are several reasons for this. They are:

1. No plan for application.
2. Techniques not made relevant to work environment.
3. Teaching techniques are out of line with the realities of the 90's.

We will address these issues in some detail.

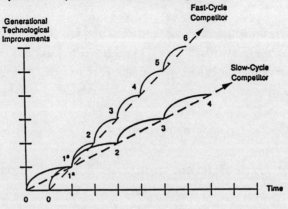

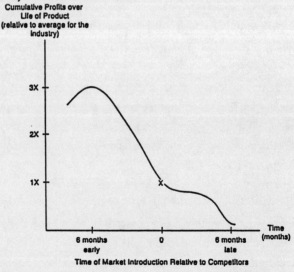

No Plan for Application

Training does not equate to behavior change. For experimental design to be applied by the masses, an environment for its use must be in place. Far too often, training is done in a vacuum. No planning is done relative to where initial applications should be done, who can provide support on applications, and what management's role is in making its application a way of life. Our recommendations in this area are:

1. Teams bring projects into the classroom.
2. Projects are discussed during the class. Potential experiments are set up.
3. Follow-up support after the session is provided.
4. Management is made aware of the importance of application and what key questions they need to ask.
5. Successful applications are rewarded.

Techniques Not Made Relevant to the Work Environment

Instructors must bring real applications into the classroom. Ideally, these applications should have been conducted on technologies familiar to the students. In addition, teams of students should meet beforehand to determine techniques to be studied, functions to be optimized, and what the possible factors and responses are.

Teaching Techniques Are Out of Line with the Realities of the 90's

We believe that there are at least seven realities that all must understand:

1. Experimental design is a strategic weapon to battle the competition on a world-wide basis by designing mandatory robust products, reducing time to market, improving quality and reliability, and reducing life cycle cost.
2. Development, design, and manufacturing must all wisely apply experimental design techniques.
3. Look for the engineering related response.
4. The application of experimental design techniques involves the blending of engineering, planning, communication, team, and statistical skills.
5. Focus on variance reduction.
6. Avoid guru worship — learn from all.
7. Use the "Texas Approach" to teach.

REALITY #1

Experimental design is a strategic weapon to battle the competition on a world-wide basis by designing mandatory robust products, reducing time to market, improving quality and reliability, and reducing life cycle cost.

A designed experiment is one where we make purposeful changes to the inputs (factors) of a process in order to observe corresponding changes in the output (responses).

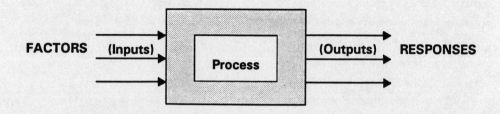

A familiar example might be:

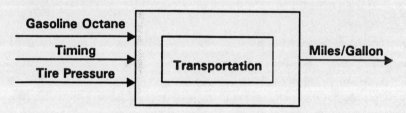

Obviously, there will be some features we will be able to control and others we can't, or simply choose not to control. Those that we won't control will be noise factors. For example:

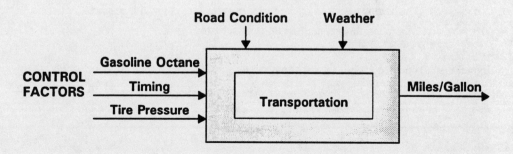

A designed experiment is **not** a substitute for chemistry, physics, and engineering. And it is most definitely **not** a substitute for good engineering judgement. As a matter of fact, the more you know about your process, the more impressive your results using these techniques!

The text *Made in America* [3] found six recurring patterns of weakness across various American companies. These common weaknesses keep United States companies from being competitive on a world-wide basis. The weaknesses are:

1. Outdated strategies
2. Short time horizons
3. Technological weaknesses in development and production
4. Neglect of human resources
5. Failures of cooperation
6. Government and industry at cross-purposes

The text continues by concluding that weakness #3 does not relate to the invention process, but rather to the **process** of turning new inventions into useful products.

The **design processes** used by American companies in the past have tended to produce designs which were overly complex and did not necessarily focus on customer requirements. What we need are design processes developing robust products and processes that are "insensitive." Insensitive means that products and processes perform their intended functions regardless of:

1. Customer imposed usage environment
2. Manufacturing variations
3. Variations imposed by vendors in subsystems and piece-parts
4. Degradation over the useful product life

Experimental and robust design techniques coupled with quality function deployment (QFD) and concurrent engineering initiatives are powerful tools used to make this a reality. For example, a fundamental requirement of QFD is to determine the strength of relationships between:

1. Customer requirements and technical requirements.
2. Technical requirements and part characteristics.
3. Part characteristics and process characteristics.

Simple relationships (particularly between customer requirements and technical requirements) may already be part of the body of engineering knowledge. Often, however, complex relationships are not understood. In these cases, experimental design techniques making use of powerful orthogonal arrays are the premier methods available today for making this a reality.

Another shortcoming of traditional American design processes was that they required excessive design times. In 1984, leading computer disk drive manufacturers frequently took up to sixty months to develop and produce new data storage devices. Today, leading companies are dictated by market constraints to do this in as little as sixteen months. Experimental design is the strategic weapon for reducing time to market by maximizing the amount of information about a product or process in a minimal amount of time.

As part of the QFD process, experimental design helps us understand the relationship between customer quality requirements and process parameters. Empirical math models relating input parameters to quality characteristics allow us to optimize quality. In turn, optimization of the quality characteristics along with robust design increases product reliability.

Engineers need to produce robust designs in minimum time and in a low cost manner. Experimental design used in the research and development phase maximizes information with minimum resources. Products and processes must be made robust **before** they go to manufacturing. Waiting to "robustize" a product/process in manufacturing is too late and not cost effective. With each life cycle phase of a product, the cost of failures increases in magnitude. Experimental design used to create robust designs with increased quality and reliability will decrease overall life cycle cost.

World-class companies in the 90's will learn to excel in these endeavors.

REALITY #2
   Development, design, and manufacturing must all wisely apply experimental design techniques.

Differing groups within your organization have different needs relative to the experiments they must conduct. Based on hundreds of experiments, we believe the following four objectives best summarize the experimental design process:

8   Straight Talk on Designing Experiments

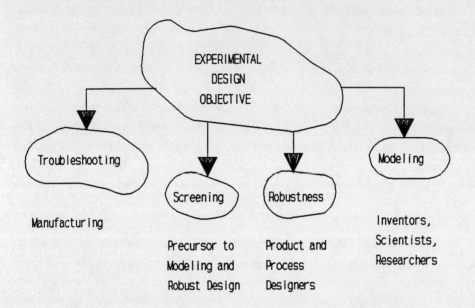

Figure 1. Functional Group Use of Experimental Design

Knowing your objective helps guide you through the first part of the experimental design process. The following examples should give you an idea of those processes we believe fall under each classification.

Manufacturing/Troubleshooting: Your process is typically in place with big problems, perhaps a line down. Your job is to find and fix the problem.

Example 1:

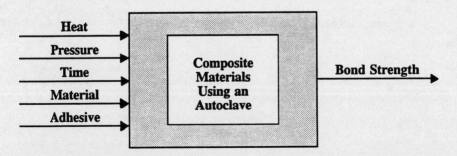

Screening: This is typically done before any of the other three when you are unsure of the important factors affecting your process. Hopefully, this will allow you to reduce the magnitude of your experiment.

Example 2:

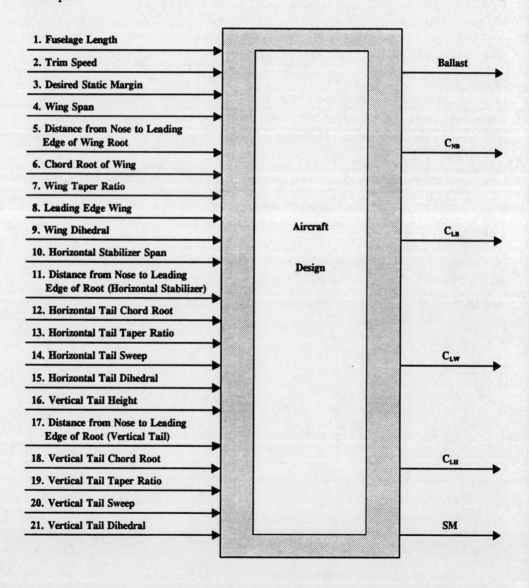

Robust Design: Our objective here is to make our process insensitive to noise. In other words, we want to design products that perform intended functions without failure regardless of varying customer usage conditions. This will allow wide variation in customer usage, allow for product deterioration, and even allow for wide variation in subsystem/component parts.

Example 3:

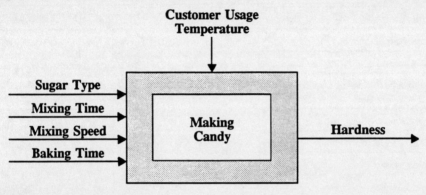

Modeling: Commonly accomplished in development or by advanced manufacturing people who need to learn a lot about a process or technology.

Example 4 [4]:

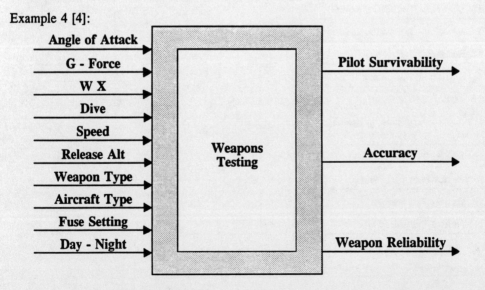

REALITY #3
Look for the engineering related response.

Consider an example of a dry film photoresist process for multilayer printed circuit boards. Possible control factors for the process include developer concentration, spray pressure, temperature, and exposure energy. Line width is a key customer requirement. The specification for line width is .006 to .007 inches. One approach would be to evaluate a yield related response, i.e. the percentage of parts meeting the specification. A better response would be to evaluate line width on a categorical scale. Those parts less than the lower specification could be placed in category "below." Those meeting specification could be placed in category "meets" and those above the upper specification could be placed in category "above." A third (and best) way is to assess the actual line width measurement and treat this as a continuous response value.

Guidelines we like to suggest to our clients in selecting responses are:

1. Pick a continuous response whenever possible.
2. Try to find a response which is easy to measure. The measurement system for the selected response needs to be precise, accurate, and stable.
3. Start with the customer requirement. Try not to stop there.
4. Select a response which relates to the engineering function of the subsystem or process. Perform a functional analysis prior to selecting responses.
5. For complex systems, attempt to break into subsystems. Do experiments at the subsystem level. Rule of thumb: For each critical engineering function, consider one experiment. Remember the old adage of "How to eat an elephant – one bite at a time."
6. Try to focus on engineering function, not problems.

An example from the photocopy business relates to Figure 2 on the following page. The function of the paper handling device in the picture is to select a sheet of paper from the stack and deliver the sheet to the paper transport system at the "nips." The vacuum acts to pick up a sheet and deliver it to the next stop. If no sheet is delivered or two sheets are delivered to the nips, then a failure has occurred.

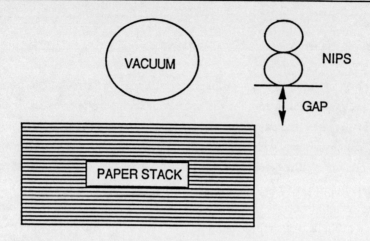

Figure 2. Side View of Paper Handling Device Schematic

Initially, you might consider the response to be whether a sheet of paper has been delivered or not (binary). A better response is one which relates to the fundamental engineering function of the subsystem. A key to meeting the customer requirement (delivering a sheet of paper to the nips) is displayed in Figure 3.

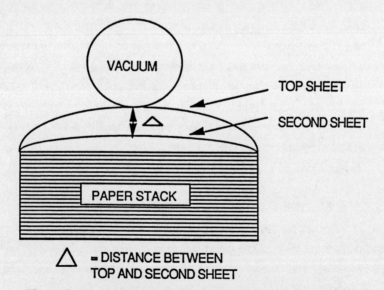

Figure 3. Front View of Paper Handling Device Schematic
(Distance Between Top and Second Sheets of Paper)

This system functions best when we are able to maximize the distance (delta) between the top and second sheet of paper in the stack. By conducting a designed experiment using this continuously measurable difference as the engineering related response, one can maximize the information using a minimum of resources.

REALITY #4
The application of experimental design techniques involves the blending of engineering, planning, communication, team, and statistical skills.

In the past, statistical skills have been overemphasized. Fifteen years ago, one of the authors attended an experimental design course at a midwest university. The course consisted of ten weeks of crunching numbers on a TI-52 calculator. Attendees became particularly adept at cranking out total sum of squares, residual sum of squares, mean square error, degrees of freedom, and F and t statistics. Unfortunately, we became competent only in plugging numbers as opposed to being able to do designed experiments. In today's environment, applying orthogonal arrays to real problems requires a diversity of skills, perhaps the least of which is being able to crunch numbers with a calculator. Technical knowledge, the ability to plan and implement, communication, and a host of people skills are essentials for success.

We like to draw the following "box and bubble" chart when discussing the industrial experimentation process:

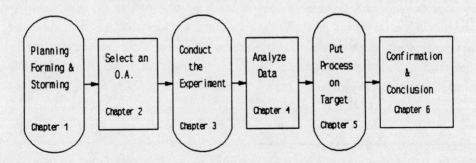

Figure 4. Experimentation Process

Note the diverse skills required for success. In the "planning" stage, primary activities which must take place are:

1. Generating a statement of the problem or opportunity.
2. Selecting an objective.
3. Determining the experimental response.
4. Identifying control factors, noise factors (if required) and number of levels.

In order to be successful in this arena, the group needs to have the following skills/knowledge.

1. Customer requirements
2. Understanding of objectives
3. Technical knowledge
4. Team skills
5. Facilitation skills
6. Communication skills
7. Resources/Time available

The second major activity, selecting an orthogonal array, was once shrouded in a high degree of statistical mysticism. In the very recent past, if an engineer wanted to conduct a designed experiment he/she would select possible factors and levels and pay a visit to the local statistical guru. The guru's office was typically in a cage with large black bars reaching to the ceiling. The engineer would quietly walk to the cage, slide the relevant information under the bars and leave abruptly as the statistician grumbled gruffly something about go away and come back in a couple of days. Employing the latent tricks of their trade, the statisticians would religiously lay out an intricate fractional factorial design. Later, the engineer would return to accept the worksheet, having no idea (or desire to learn) how the runs were generated.

No longer must each design be generated by hand. Thanks to the common availability of tabled orthogonal arrays and easy to use software, the engineer can quickly be guided in selecting the proper orthogonal array to fit the problem.

The third stage, conducting an experiment, requires:

1. A detailed plan
2. Discipline in following the plan
3. An understanding by all of what the experiment is to accomplish.

Most texts on design of experiments give little or no press to this stage. It is, however, one of the biggest reasons why people obtain poor results from an experiment. Those involved must realize the only changes to be made are those called for in the orthogonal array. All other potential sources of variation must be held constant, or as close to constant, as possible.

Conducting analysis is still seen by some, unfortunately, as what experimental design is all about. For example, many courses and texts on design of experiments may spend as much as 80-90% of their time in this arena. Analysis of variance and regression are typically touted as the techniques of choice. Recent revelations, however, have shown us that with simple graphs and sixth grade math, we can obtain a solid fundamental understanding of orthogonal array experiments. Analysis of variance and regression output tables, generated by the computer, can provide additional information from the data.

The last major block, confirmation and conclusions, is the "proof in the pudding" step. As the name implies, we take the predicted best settings from the analysis stage and run those combinations with 4-20 repetitions (depending upon cost). If the results of the confirmation closely match the predicted values, our experiment has confirmed. If not, this indicates one or more of our assumptions is not valid. Failure to confirm directs us to attempt to find reasons for the unpredictability prior to initiating new experimental trials.

The reality of using engineering, planning, communication, teamwork, and statistical skills throughout the experimental design phases will lead to more successful results.

REALITY #5
Focus on variance reduction.

Until the recent past everyone assumed variability was constant over the entire experimental region. Designed experiments focused upon determining those factors which shifted the average. A new paradigm has emerged in just the last few years.

Engineers are now learning there are not only factors which shift the average, but factors which impinge upon the variation as well.

An enlightening study we recently undertook with a client helps highlight this point. Because of the proprietary nature of the process used in the production of this part, we can only say the following:

1. The part consists of a piece of screen (looks like screen door screen) approximately 3" in length by 1/2" wide. The screen is wrapped around a core and welded together at several locations.
2. The diameter of the core is critical.
3. Recent process capability studies indicate the process is not capable of meeting specifications. In fact, 100% audits of the process have been initiated resulting in approximately $200,000 of scrap and rework costs being incurred.

A simple screening experiment was set up and run. The orthogonal array and coded results are shown in Table 1:

Table 1. Experimental Design Orthogonal Array and Coded Results

Run	Control Factors							Target=797.5 Spec =785-810						
	A	B	D	C	E	F	G	Diameters					Avg	St Dev
1	−	−	−	−	−	−	−	806	805	809	806	804	806	1.87
2	−	−	−	+	+	+	+	804	802	804	805	801	803	1.64
3	−	+	+	−	−	+	+	803	808	811	816	806	809	4.97
4	−	+	+	+	+	−	−	812	806	813	802	804	807	4.88
5	+	−	+	−	+	−	+	796	805	805	810	794	802	6.75
6	+	−	+	+	−	+	−	799	806	801	803	811	804	4.69
7	+	+	−	−	+	+	−	802	804	796	804	800	801	3.34
8	+	+	−	+	−	−	+	803	802	805	804	805	804	1.30

A, B, C, D, E, F, and G are control factors set at two different levels ("−" is one setting and "+" is a second setting). As shown, eight recipes or runs were conducted. Five parts were produced for each run with the applicable outer diameters measured. An average and standard deviation of the five diameters for each of the eight runs were calculated.

For demonstration analysis, Table 2 compares column "D" to the standard deviation(s) column:

Table 2. Comparison of Factor D to the Standard Deviation

Run	Column D	Standard Deviation
1	−	1.87
2	−	1.64
3	+	4.97
4	+	4.88
5	+	6.75
6	+	4.69
7	−	3.34
8	−	1.30

Note that the standard deviation(s) appear much larger when D is at the "+" level than at the "−" level. Additional analysis can be used to substantiate that D is in fact one of those wonderful factors which impinges upon the variation. By setting factor D at the "−" setting, our client was able to dramatically reduce the variation in the response diameter. Pictorially, the improvement is shown in Figure 5.

After implementing the improved process, scrap and rework were dramatically reduced.

By simply selecting the best setting for factor D, dramatic improvements were displayed in the process. **Note:** Few people have learned to use experimental design to reduce variation. Leading companies will learn to make use of this Reality.

18 Straight Talk on Designing Experiments

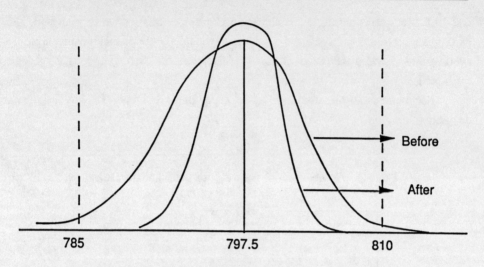

Figure 5. Diameter Variation Reduction of Part

REALITY #6
  Avoid guru worship — Learn from all.

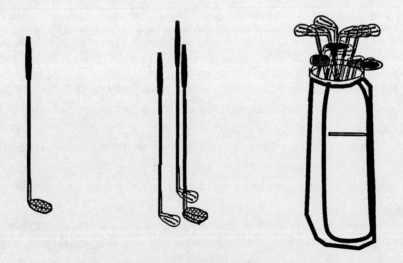

Figure 6. Are you playing with a full set of clubs?

Suppose you are playing in the company golf tournament. First prize is a $500 gift certificate plus an opportunity to socialize with executive management. Which set of clubs would you select? Will you select the set with just the driver, the set with the driver, nine iron, and five iron, or the complete set? The full set makes a great deal of sense to us.

The guru's of design of experiments all have some wonderful ideas. Unfortunately, however, there is only one way to do things — their way (they want you to use only one or three clubs). It is important to study the guru's, but not to become a "groupie" of just one. Learn from them all. Take the best from each of the competing strategies, then use what works for you. Your organization is faced with a diversity of challenges. Becoming world class requires a full set of clubs.

REALITY #7
Use the "Texas Approach" to teach.*

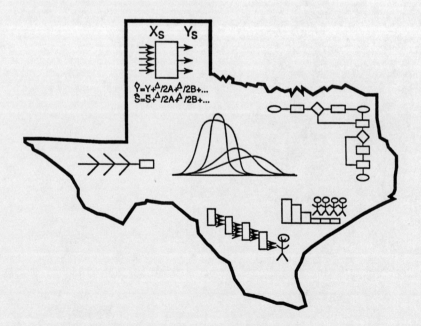

Figure 7. Texas Approach to Teaching Experimental Design

---

* The idea for the Texas Approach evolved from conversations with Dr. S. R. Schmidt.

In Texas, folks like their jokes about Aggies, tea-sippers, bubbas, and yuppies. They expect ya'll to cut out the "bull" and "cut to the chase." They do not care much about who invented an approach as long as it gets results. Think about Texas when teaching design of experiments. Tell some jokes. Show some pictures (as in Figure 7). Design of experiments can be down to earth if we teach it that way. If we "cut out the bull," we do not need to spend days in the classroom generating fractional factorials and analyzing data using matrix algebra.

Engineers and scientists have no desire to be statisticians. They just want an easy-to-use, practical book. Let us give it to them. Use pictures to help explain the results. Use computers for the math. Let the scientists and engineers focus on what they care about—their technologies and how design of experiments can be used as a powerful tool in helping them better understand their technology.

Introduction Bibliography

1. Stalk, George Jr. and Hout, Thomas M., *Competing Against Time*. New York: The Free Press, 1990.

2. Wheelwright, S. C, and Clark, K. B., *Revolutionizing Product Development*. New York: Macmillian, Inc., 1992.

3. Dertouzos, Michael L; Lester, Richard K.; and Solow, Robert M., *Made in America*. Cambridge, MA: MIT Press, 1989.

4. Participant's Guide for Four-Day Design of Experiments Course, Schmidt/Launsby Consulting.

II. The Basics

We present the basics of experimental design with the "box and bubble" chart we introduced as the fourth reality of the nineties (page 13):

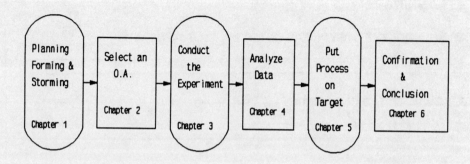

This chart provides a logical framework to guide the beginning experimenter through his or her first experimental design. Chapters 1-6 cover each of these bubbles and boxes in greater detail and relate each to one example, a small wooden catapult that we use in our seminars. This gives us the opportunity to demonstrate the power of experimental design with a process most of us understood in elementary school.

Because the boxes and bubbles don't give us much detail, we have modified a form from our Participant's Guide and included it in the next three pages. You may want to follow this form closely as you complete your first few designed experiments.

STEPS IN CONDUCTING EXPERIMENTS

I. PLANNING

A. WHO ARE THE CUSTOMERS (INTERNAL OR EXTERNAL)? _____
_____
_____

B. HOW WILL THE CUSTOMERS USE THE PRODUCT (CUSTOMER INTENT)?
_____
_____

C. WHAT FUNCTIONS MUST THE PRODUCT PERFORM TO MEET CUSTOMER INTENT? _____
_____

    1. Perform a functional analysis. _____
    2. Determine which functions require further study. _____
_____

    NOTE: 1. For each function you wish to optimize, consider one designed experiment.
            2. When appropriate, diagram the function.

D. OBJECTIVE OF THE EXPERIMENT: _____
_____
_____

E. START DATE: _____ **END DATE:** _____

F. SELECT QUALITY CHARACTERISTICS (also known as responses, dependent variables, or output variables). These characteristics should be related to customer needs and expectations.

	RESPONSE	TYPE	How will you measure the response? Is the measurement method accurate and precise?
1.			
2.			
3.			

STEPS IN CONDUCTING EXPERIMENTS (CONT)

G. SELECT FACTORS (also known as parameters or input variables) which are anticipated to have an effect on the response. _____
_____

FACTOR	TYPE	CONTROLLABLE OR NOISE	RANGE OF INTEREST	LEVELS	ANTICIPATED INTERACTIONS WITH	HOW MEASURED
1.						
2.						
3.						
4.						
5.						

H. DETERMINE THE NUMBER OF RESOURCES TO BE USED IN THE EXPERIMENT. (Consider the desired number, the cost per resource, time per experimental trial, and the maximum allowable number of resources.) _____
_____
_____

II. SELECT AN ORTHOGONAL ARRAY

A. SELECT THE BEST DESIGN TYPE AND ANALYSIS STRATEGY TO SUIT YOUR NEEDS. _____
_____

B. CAN ALL THE RUNS BE RANDOMIZED? _____
WHICH FACTORS ARE MOST DIFFICULT TO RANDOMIZE? _____
_____

III. CONDUCT THE EXPERIMENT AND RECORD THE DATA. (Monitor both of these events for accuracy.)

STEPS IN CONDUCTING EXPERIMENTS (CONT)

IV. ANALYZE THE DATA

A. DETERMINE IMPORTANT EFFECTS. _____

B. DETERMINE BEST SETTINGS FOR FACTORS. _____

C. GENERATE A PREDICTION INTERVAL. _____

D. PUT RESPONSE ON TARGET WITH MINIMUM VARIATION. _____

V. CONFIRMATION

A. CONDUCT 4-20 (OR MORE) TESTS AT THE "BEST SETTINGS." _____

B. COMPARE RESULTS TO BENCHMARK AND PREDICTION INTERVAL. _____

C. ASSESS RESULTS AND PLAN THE NEXT STEPS. _____

CHAPTER ONE
PLANNING
(The First Eight Steps)

Let's begin with the first "bubble" on our chart. A tool that we commonly use in our classes is a catapult, affectionately referred to as a "statapult." Figure 1.1 illustrates the basic setup:

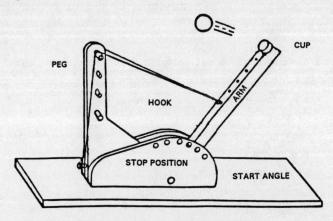

Figure 1.1 The Statapult

There are five obvious adjustments:

1. Pull back angle of the arm (Pull back angle)
2. Stop position (Stop)
3. Peg on upright arm for rubber band tension adjustment (Peg)
4. Adjustable rubber band connector to arm (Hook)
5. Cup position (Cup)

As we move through the remaining chapters, we will refer to the statapult many times. (If you are interested, statapults may be purchased through Launsby Consulting 1-800-788-4363.)

A. **Who are the customers?** Although this may seem self-explanatory, do not forget to consider both **internal and external** customers. Internal customers are those we have some control over. External ones are those we do not. If we made the statapult on an assembly line, the workers who install the pins or stops would be customers of those who drill the holes. The external customers might be those who instruct the course using the statapult, as well as students who use it in class.

B. **How will the customers use the product (intent)?** As an instructional tool, we will probably just use the statapult to fire rubber balls. As instructors, we take four of these with us to every class. Being able to dismantle them gives us a little more "suitcase space." However, they must be able to weather the frequent reassemblies as well as the travel.

C. **What functions must the product perform to meet customer intent?**
   1. Perform a functional analysis. A function is defined as that which makes a product **work** or **sell**. The following rules will help us determine applicable functions:
       a. The expression of all functions should be accomplished in two words — a verb and a noun.
       b. Separate work and sell functions.
       c. The expression of work and sell functions uses different categories of verbs and nouns:
           (1) Work functions are usually expressed in action verbs and measurable nouns which establish quantitative statements:

VERBS	NOUNS
support, change	weight, force
transmit, interrupt	light, oxidation
create, establish	heat, flow
hold, shield, emit	radiation, current
enclose, modulate	friction, insulation
collect, control	voltage, energy
conduct, insulate	force, density
protect, repel	damage, circuit
prevent, filter	
reduce, impede	

(2) Sell functions are usually expressed in passive verbs and non-measurable nouns which establish qualitative statements:

VERBS	NOUNS
increase	beauty, symmetry
decrease	appearance, effect
improve	convenience, exchange
	style, features, form

To accomplish a functional analysis, we must essentially dissect our product. The pieces may suggest responses or factors we would have otherwise forgotten. For the statapult, the following "exploded" view aids in completing the functional definition (Table 1.1).

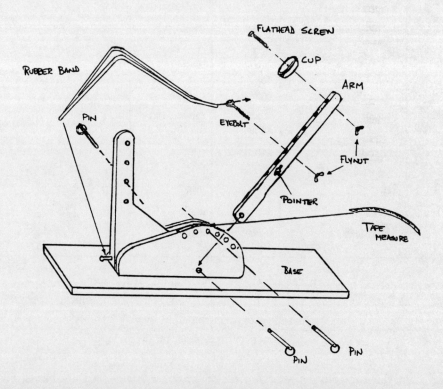

Figure 1.1a Exploded View of the Statapult

Table 1.1 Functional Definition of the Statapult

FUNCTIONAL DEFINITION				
PROJECT: STATAPULT				
INPUT _ball, force_		BASIC FUNCTION _launch ball_		OUTPUT _distance, altitude_
QTY	PART	FUNCTION(S)		COMMENTS
		VERB	NOUN	
1	Arm	throw	ball	
1	Cup	hold	ball	
1	Statapult Base (includes stop position & upright arm)	stop hold attach	arm pins rubberband	
1	Rubber Band	create	force	
3	Pin	stop adjust hold create	arm tension arm fulcrum	
1	Eyebolt	attach adjust	rubberband tension	
2	Fly Nut	secure secure	eyebolt cup	
1	Ball	hit	target	
1	Pointer	set	angle	
1	Tape Measure	measure	distance	

(3) Determine which functions require further study.
    a. For each function you wish to optimize, consider one designed experiment.
    b. When appropriate, diagram the function.

For our statapult, we want to consider the entire unit as the most basic, with the function "launch ball."

D. **Objective of the Experiment:** In terms of DOE, you will be more successful if you decide which DOE objective you are trying to accomplish:

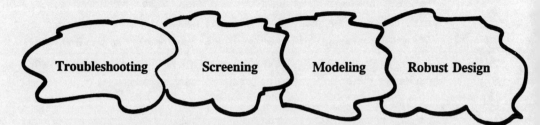

We will recommend different experimental designs for each objective. In class we illustrate all four objectives with the statapult:

Day 1: Accomplish a screening experiment to determine the three most important factors affecting downrange distance.

Day 2: Using the three factors from day 1, perform a troubleshooting experiment to put a ball on target.

Day 3: Make the statapult robust to ball type.

Day 4: Model the process as a nonlinear relationship between factors and in-flight distance.

E. **Start/End Dates:** Simply frames the time you have to conduct your experiment, analyze and interpret your results, and ready presentations for management.

F. **Select Quality Characteristics:** Your experiment will be driven by the process output (response) that you choose to observe. Of course, meaningful and measurable responses will help us narrow our process inputs (factors). As suggested in the February 1992 *Quality Digest*, here are some guidelines to selecting responses:

- Select a response that is measurable on a continuous basis. Categorical responses such as "pass/fail" require much larger sample sizes. (Rule of thumb is $N \times p \geq 5.0$, where "N" is the number of response values and "p" is the proportion defective. For example let's suppose we are trying to improve the paper feed mechanism on the copier mentioned in the introduction. If one copy in ten ($p = 1/10$) jams, then we would need to test at least 50 samples. However, if only one copy in one hundred ($p = 1/100$) jams, the minimum sample size jumps to 500. The measurement device for the response must be precise, accurate, and stable.

- Select an engineering-related response(s) that has a cause-and-effect relationship with the customer requirement. ("Customer" may be an internal user as well as the ultimate user.) This will not always be easy to do. **Use of your engineering knowledge** and failure analysis techniques can be of benefit.

- Try to use responses that are easy to measure.

- When possible, use responses that are not heavily influenced by interactions.

- If your problem relates to complex products or processes, try to break your experiment down into the major steps or subsystems. The answer to "How do you eat an elephant?" applies in this case.

The best response will be a continuous one (e.g., downrange distance). The worst response is a categorical one (e.g., hit target/miss target). Responses of more than two categories fall somewhere in between. If visual inspections are required, we recommend you limit categories to three to five. In any case, your measurement system is critical — it should be accurate and precise! At the very least, it must be stable.

For our statapult experiment, we decided to use in-flight distance as our response. If customer feedback indicated other subsystem problems (e.g., ball sticks in the cup, rubber band breaks after six shots, statapult does not sit flat, pins do not fit in drilled holes, etc.), we would attack each with a separately designed experiment. To measure in-flight distance, place a measuring tape outward from the end of the statapult base and have two spotters mark the ball impact point. Since the distance is measured at any point between zero and the end of the tape measure, this response is continuous. Assuming our customers want to hit specific targets, in-flight has a cause-and-effect relationship with their requirement. We believe this response is easy to measure, although there may be induced variability with human spotters. We expect interactions to influence the distance, but we are not sure how significantly.

G. **Select Factors**: Recall that we are selecting factors that we can purposefully change to observe corresponding changes in the response(s). Combining your technical knowledge with your functional analysis, charts such as the cause-and-effect diagram (fishbone) may be useful in your brainstorming session (Figure 1.2).

By assembling the factors that affect downrange distance, it should not take long to identify those we can control and those we cannot (or choose not to). Recall that those factors we can control are called control factors and those we cannot or choose not to control are called noise factors. For our troubleshooting example, we will ignore the noise factors.

DOWNRANGE DISTANCE

Control Factors	Noise Factors
Pull Back Angle	Air Flow in Room
Stop Position	Ball Weight
Peg on Upright Arm	
Hook on Pullback Arm	
Cup Position	
Spotter	
"Archer"	

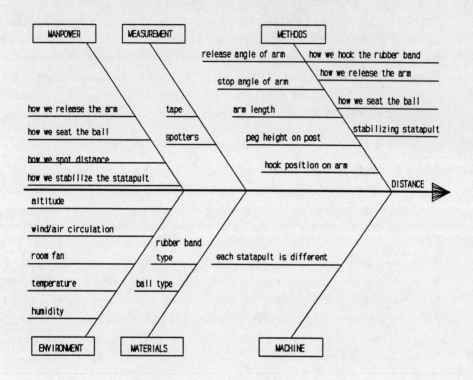

Figure 1.2 Cause-and-Effect Diagram of Statapult Experiment

As we did with responses, we prefer factors that can be varied on a continuous scale. This may not always be possible, as our statapult demonstrates. Only the pullback angle is a continuous factor. We will also need to select high (+) level, low (−) level, and possibly mid level (0) settings for our factors. If we suspect our response reacts linearly to a factor, then two levels are probably adequate (Figure 1.3). But if our technical insight tells us that the relationship between the factor and the response is probably "curved" or nonlinear, then three or more levels would be more appropriate (Figure 1.4).

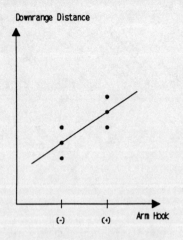

Figure 1.3 Linear Relationship between Factor and Response

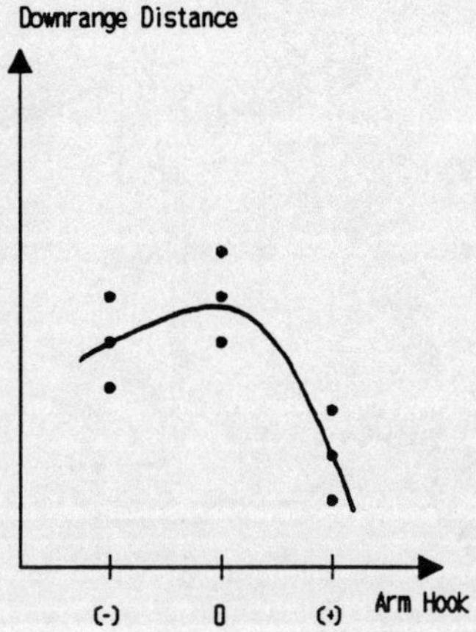

Figure 1.4 Nonlinear Relationship between Factor and Response

The low and high settings we decided upon were:

Factor Levels

Factors	Low Level	High Level
Pull back angle	Full back	1/2 back
Stop	Hole 1	Hole 3
Peg	Lowest hole	Highest hole
Hook	Lowest hole pull back	Third hole down
Cup	Top hole pull back	Second hole down
Spotter	Spotter 1	Spotter 2
"Archer"	Archer 1	Archer 2

Figure 1.5 shows the statapult settings:

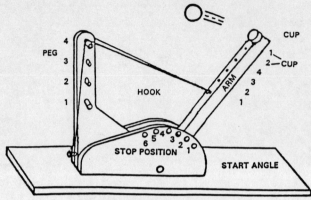

Figure 1.5 Statapult Settings

For a nonlinear process, you must also select a mid-level setting that falls mid-way between your low and high settings. These should be attainable in your process. If you intend to model a nonlinear process using more than three levels (Chapter 11), you must leave room for settings that are above and below those high and low settings. More than three levels are not normally required.

Note the importance of range of interest in Figure 1.4. If we assume only a linear relationship between stop position and in-flight distance and experiment at only two of these levels, our process model will be very inaccurate at the third level (Figure 1.6).

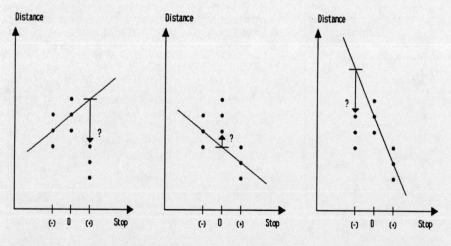

Figure 1.6 In-Flight Distance vs. Stop Position (Inappropriate Linearization)

However, if we are only interested in modeling our process from the low to mid-levels or from the mid to high levels, a linear approximation may be more than adequate (Figure 1.7).

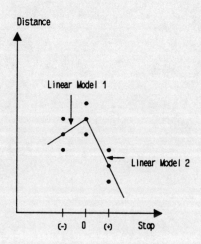

Figure 1.7 In-Flight Distance vs. Stop Position
(Appropriate Linear Approximation)

We use statapult factor settings that span a linear range of responses to illustrate a two-level experiment.

H. Determine the number of resources to be used in the experiment. Cost will be an obvious driver for most experiments, along with the time to complete the experiment. We normally limit our students to 30 shots and 2 1/2 hours.

The remainder of the items on our planning forms are the subjects of the next five chapters.

CHAPTER TWO
SELECTING AN ORTHOGONAL ARRAY

> A. Select the Best Design Types and Analysis Strategy to Suit Your Needs

2.1 Design Matrix

A design type or design matrix (array) is a way to organize and track an experiment. Let's suppose our brainstorming session or a screening experiment suggests that pull back angle, hook, and peg have the biggest influence on in-flight distance. We would like to conduct an experiment with the three factors at different settings and record the in-flight distance for each combination. We might organize this in a table as:

FACTORS			RESPONSE
Pull Back Angle	Hook	Peg	Distance

If our technical knowledge suggests the relationship between the factors and the response is linear, we may elect to look at only two levels of each factor — a low and a high level. Based on Figure 2.1, the factors and the corresponding ranges for their levels might be:

        Pull Back   (full back, 1/2 back)
        Hook   (1, 5)
        Peg   (1, 4)

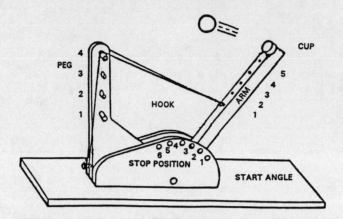

Figure 2.1 Statapult Factor Levels

The tree diagram in Figure 2.2 illustrates the number of runs required if we were to experiment at all possible combinations of the factor settings.

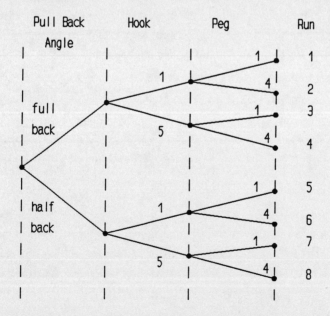

Figure 2.2 Tree Diagram for Three Factors at Two Levels

The diagram clearly shows that we would have to run eight [(number of levels)$^{\text{(number of factors)}}$ = 2^3 = 8] different combinations of these factors settings. Placing these eight combinations in the design matrix we have:

Run	Pull Back	Hook	Peg	Distance
1	Full back	1	1	
2	Full back	1	4	
3	Full back	5	1	
4	Full back	5	4	
5	1/2 back	1	1	
6	1/2 back	1	4	
7	1/2 back	5	1	
8	1/2 back	5	4	

Each combination is known as a "run." In this "orthogonal" matrix, we will arbitrarily assign the low value a -1 and the high value a $+1$ (a mid-value would be 0). Therefore, the previous design matrix can be rewritten with -1 and $+1$ shortened to "$-$" and "$+$."

Run	Pull Back	Hook	Peg	Distance
1	−	−	−	
2	−	−	+	
3	−	+	−	
4	−	+	+	
5	+	−	−	
6	+	−	+	
7	+	+	−	
8	+	+	+	

2.2 Orthogonal Designs

Mathematics introduces orthogonality as a measure of perpendicularity. Linear algebra goes a little further and speaks of the dot product of two vectors equaling zero when they are orthogonal. Generally, orthogonality describes the independence among vectors, factors, functions, etc. Part of our goal is to determine the effect each of the factors has on the response and on the variation of the response, mathematically independent of the effects of the other factors.

The matrix we have just built is orthogonal. It is both vertically and horizontally balanced. For each factor, we will test at an equal number of high and low values (vertical balancing). For each level within each factor, we are testing an equal number of high and low values from each of the other factors (horizontal balancing). (This definition of an orthogonal design will suffice for now. The precise definition of an orthogonal matrix will be discussed in Chapter 4.)

Mathematically speaking, vertical balancing occurs if the sum of each column is zero. Horizontal balancing is evaluated much like a vector dot product. If the sum of the products of corresponding rows in two columns is zero, those two columns are orthogonal. The design is horizontally balanced if each two column combination sums to zero.

Run	Pull Back	Hook	Peg	Pull Back × Hook	Pull Back × Peg	Hook × Peg	Distance
1	−	−	−	+	+	+	
2	−	−	+	+	−	−	
3	−	+	−	−	+	−	
4	−	+	+	−	−	+	
5	+	−	−	−	−	+	
6	+	−	+	−	+	−	
7	+	+	−	+	−	−	
8	+	+	+	+	+	+	
Sum (Σ)	0	0	0	0	0	0	

If we test all possible combinations of our factors, our design matrix will be referred to as a **full factorial**. This will always be one of our design options.

Notice that in the illustration of horizontal balancing, we have created three new columns in the design matrix. These will be used to examine possible interactions of factors. Orthogonal designs even ensure horizontal balancing between interactions and factors. We will see in Chapter 5 that these types of designs also simplify our analysis.

We may decide that a full factorial may require too many resources, or that a slightly nonorthogonal array may be acceptable. Let's begin this examination with the array we developed above with another column added for a possible three-factor interaction.

Run	Pull Back	Hook	Peg	Pull Back × Hook	Pull Back × Peg	Hook × Peg	Pull Back × Hook × Peg	In-Flight Distance
1	−	−	−	+	+	+		
2	−	−	+	+	−	−		
3	−	+	−	−	+	−		
4	−	+	+	−	−	+		
5	+	−	−	−	−	+		
6	+	−	+	−	+	−		
7	+	+	−	+	−	−		
8	+	+	+	+	+	+		

2.2.1 Fractional Factorials and Aliasing

On the first day of our four day seminar we typically have our students do a screening experiment with the seven factors we introduced in Chapter 1. Again, the following tree diagram (Figure 2.3) gives you a picture of the number of runs required to test all possible combinations of the factor settings:

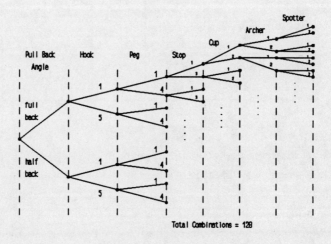

Figure 2.3 Diagram for Seven, Two-Level Factors

Because we limit them to eight runs, they must somehow accomplish a valid experiment with a reduced design matrix. We saw in Figure 2.3 that all possible combinations of factor settings requires $2^7 = 128$ runs. With the above array, it is possible for us to take a look at all seven main factors and identify most significant ones. If our technical knowledge suggests that interactions have relatively small effects, we can replace those column headings with the remaining four factors:

> Stop
> Cup
> Archer
> Spotter

The design matrix now looks like:

Run	Pull Back	Hook	Peg	STOP Pull Back × Hook	CUP Pull Back × Peg	ARCHER Hook × Peg	SPOTTER Pull Back × Hook × Peg	In-Flight Distance
1	−	−	−	+	+	+		
2	−	−	+	+	−	−		
3	−	+	−	−	+	−		
4	−	+	+	−	−	+		
5	+	−	−	−	−	+		
6	+	−	+	−	+	−		
7	+	+	−	+	−	−		
8	+	+	+	+	+	+		

This does not mean the new column names are in any way equal to the previous headings. It simply means we will take advantage of the orthogonal design settings for those trivial interactions. When we hide interactions, we refer to this as aliasing. Partial aliasing is known as confounding [1].

The type of design we now have is referred to as a **fractional factorial**. For our seven factor, full factorial we require $2^7 = 128$ runs. Since we aliased four interactions with main factors, we have a $2^{7-4} = 2^3$ design. Texts list these as 2^{k-q} fractional factorials (k = # of factors, q = # of factors assigned to interaction columns when generating the design). Note that the fractional factorial is also orthogonal.

2.2.2 Resolution

One method of measuring aliasing in two-level designs is called resolution. Many of the available software packages provide the resolution for a selected design. Resolution may be measured II or higher, with the higher number indicating less significant aliasing. Confounding is measured with a correlation coefficient and will be discussed later.

Resolution Chart − Definitions

R(V): Main effects and 2-way or 3-way interactions are not aliased with each other. (Unsaturated design) (i.e. New Factor = Pull Back × Hook × Peg × Spotter)

R(IV): Main effects are aliased with 3-way interactions, 2-way interactions are aliased with other 2-way interactions. (Unsaturated design) (i.e. Spotter = Pull Back × Hook × Peg)

R(III): Main effects are aliased with 2-way interactions. (Saturated design) (i.e. Stop = Pull Back × Hook)

R(II): Main effects are aliased with other main effects. (Supersaturated design) (i.e. Pull Back = Stop)

The fractional factorial above is a resolution III design since main effects are aliased with two-factor interactions. Since it is typical for the effects of factors to be much larger than the effects of interactions, this type of design is useful for screening out insignificant factors with large effects.

2.2.3 Tabled Taguchi* Designs

Table 2.1 will help with your strategy on aliasing interactions. This table

---

* Dr. Genichi Taguchi is credited with much of the practical use of Experimental Design techniques. A native of Japan, Dr. Taguchi is best known for this development of signal-to-noise ratios while working as a mechanical engineer at Nippon Telephone and Telegraph. These ratios are not those used in electrical engineering. Dr. Taguchi gave his statistic involving target responses with minimum variation this name to encourage its use among co-workers − mostly electrical engineers.

summarizes a set of the seven most frequently used Taguchi designs. These designs were modified from full and fractional factorials, Plackett-Burman designs, Hadamard matrices, and Latin Square designs.

Table 2.1 Most Frequently Used Taguchi Designs [1]

Design	# of Levels	# of Factors for the Full Factorial	# of Factors to Maintain Resolution V	# of Factors for Screening
L_4	2	2	2	3
L_8	2	3	3	7
L_9	3	2	–	4
L_{12}	2	–	–	11
L_{16}	2	4	5	15
L_{18}	mixed	–	–	8
L_{27}	3	3	–	13

To use Table 2.1 you must:

1. Determine your factors and corresponding levels.
    a. If your technical experience indicates linear relationships between factors and responses, then a two-level design is appropriate.
    b. If you suspect nonlinear relationships, a three-level design will better suit your needs.
    c. If you have mixed relationships in your process, the L_{18} is the only design in this table that may help (Dr. Taguchi attributes the "L" designation to his use of the Latin Squares generating matrix in his designs).

2. Decide how many runs you can afford. The "L" subscript gives you the number of runs in a particular design matrix.

3. Use columns 3-5 of Table 2.1 to develop your strategy on interactions.

Tables 2.2 - 2.8 present each of the designs listed in Table 2.1, along with a dialogue box to guide you on interactions. These designs are not all inclusive and we will discuss other useful designs following Tables 2.2 - 2.8. For example, let's consider Table 2.2:

Table 2.2 L₄ Design [1] for 2-Level Factors

	L₄ Design		
Run #	1	2	3
1	−1	−1	−1
2	−1	+1	+1
3	+1	−1	+1
4	+1	+1	−1
Original factors and interactions used to generate the matrix	a	b	−ab

Strategy	# Factors	Which Columns?
Full Factorial	2	1, 2
Resolution V	2	1, 2
Screening	3	1, 2, 3

Your first question may be about all the negative signs in the first row. Dr. Taguchi likes to assign the −1 orthogonal value to all of his low cost settings and place them in the first row. If an experiment shows good results after only one run, he may recommend the process be set at those levels for immediate improvement. To get all negatives in the first row of the design, all columns with positive values in the first row must be negated. In this case, the third column was generated not as ab, but as −ab. If you are interested in all main factors and the two-factor interaction, you should use columns 1 and 2 for your main factors. To maintain high resolution (V), you must also assign your two main factors to columns 1 and 2. For screening, you may use all three columns for main factors. This will give you a Resolution III design. To maintain the highest resolution for screening, you must alias interaction columns in the order given in the box below each tabled design. Table 2.3 clearly shows that highest resolution will be maintained in screening if you assign main factors sequentially to columns 1, 2, 4, and 7 before all others. The rest may be assigned in any order. Here, we are recommending that you alias a three-factor interaction with a main factor (Resolution IV) before aliasing two-factor interactions with other main factors (Resolution III).

Table 2.3 L_8 Design [1] for 2-Level Factors

Run #	L_8 Design						
	1	2	3	4	5	6	7
1	−1	−1	−1	−1	−1	−1	−1
2	−1	−1	−1	+1	+1	+1	+1
3	−1	+1	+1	−1	−1	+1	+1
4	−1	+1	+1	+1	+1	−1	−1
5	+1	−1	+1	−1	+1	−1	+1
6	+1	−1	+1	+1	−1	+1	−1
7	+1	+1	−1	−1	+1	+1	−1
8	+1	+1	−1	+1	−1	−1	+1
Original factors and interactions used to generate the matrix	a	b	−ab	c	−ac	−bc	abc

Strategy	# Factors	Which Columns?
Full Factorial	3	1, 2, 4
Resolution V	3	1, 2, 4
Screening	4-7	1, 2, 4, 7, ...

This is where we introduce the "Pareto Chart of the World" as shown in Figure 2.4 on the following page. This chart summarizes our experiences in dealing with interactions. Most process responses are affected primarily by main factors, fewer by two-factor interactions, and almost none are affected by higher order interactions. If you believe this view of the world, it should make sense to alias higher order interactions. (A notable exception to this view is the chemical and biotechnology industries where two and three factor interactions are often "heavy hitters.") The horizontal lines on the chart indicate how the classical and Taguchi experimental designers handle interactions. The classical view of DOE is to examine main factors and some interactions (fractional factorials). Dr. Taguchi believes heavily in our "Pareto Chart of the World" and teaches his followers to alias all interactions. Remember, most experiments boil down to time and money. If you do not believe an interaction is important, use the space for a main factor! The

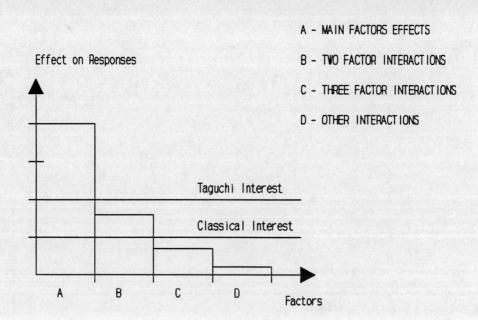

Figure 2.4 Pareto Chart of the World

decision on what interactions to alias rests on your shoulders and it relies on the original objective of your experiment — Screening, Troubleshooting, Robust Design, or Modeling. The more accurately you want to predict a response, the fewer interactions you will want to alias.

The remaining five Taguchi tabled designs appear on the following pages:

Table 2.4 L₉ Design [1] for 3-Level Factors

Run #	L₉ Design			
	1	2	3	4
1	−1	−1	−1	−1
2	−1	0	0	0
3	−1	+1	+1	+1
4	0	−1	0	+1
5	0	0	+1	−1
6	0	+1	−1	0
7	+1	−1	+1	0
8	+1	0	−1	+1
9	+1	+1	0	−1
Original factors and interactions used to generate the matrix	a	b	a+b	2a+b

Strategy	# Factors	Which Columns?
Full Factorial	2	1, 2
Screening	3-4	1, 2, 3, ...

Table 2.5 L_{12} Design [1] for 2-Level Factors

Run #	\multicolumn{11}{c}{L_{12} Design}										
	1	2	3	4	5	6	7	8	9	10	11
1	−1	−1	−1	−1	−1	−1	−1	−1	−1	−1	−1
2	−1	−1	−1	−1	−1	+1	+1	+1	+1	+1	+1
3	−1	−1	+1	+1	+1	−1	−1	−1	+1	+1	+1
4	−1	+1	−1	+1	+1	−1	+1	+1	−1	−1	+1
5	−1	+1	+1	−1	+1	+1	−1	+1	−1	+1	−1
6	−1	+1	+1	+1	−1	+1	+1	−1	+1	−1	−1
7	+1	−1	+1	+1	−1	−1	+1	+1	−1	+1	−1
8	+1	−1	+1	−1	+1	+1	+1	−1	−1	−1	+1
9	+1	−1	−1	+1	+1	+1	−1	+1	+1	−1	−1
10	+1	+1	+1	−1	−1	−1	−1	+1	+1	−1	+1
11	+1	+1	−1	+1	−1	+1	−1	−1	−1	+1	+1
12	+1	+1	−1	−1	+1	−1	+1	−1	+1	+1	−1
	Strategy				# Factors				Which Columns?		
	Screening				8-11				any		

(This is the same matrix as the Plackett-Burman, 12-run design to be discussed later; however, the columns and rows have been permuted.)

Table 2.6 L_{16} Design [1] for 2-Level Factors

L_{16} Design

Run #	1	2	3	4	5	6	7	8	9	10	11	12	13	14	15
1	−1	−1	−1	−1	−1	−1	−1	−1	−1	−1	−1	−1	−1	−1	−1
2	−1	−1	−1	−1	−1	−1	−1	+1	+1	+1	+1	+1	+1	+1	+1
3	−1	−1	−1	+1	+1	+1	+1	−1	−1	−1	−1	+1	+1	+1	+1
4	−1	−1	−1	+1	+1	+1	+1	+1	+1	+1	+1	−1	−1	−1	−1
5	−1	+1	+1	−1	−1	+1	+1	−1	−1	+1	+1	−1	−1	+1	+1
6	−1	+1	+1	−1	−1	+1	+1	+1	+1	−1	−1	+1	+1	−1	−1
7	−1	+1	+1	+1	+1	−1	−1	−1	−1	+1	+1	+1	+1	−1	−1
8	−1	+1	+1	+1	+1	−1	−1	+1	+1	−1	−1	−1	−1	+1	+1
9	+1	−1	+1	−1	+1	−1	+1	−1	+1	−1	+1	−1	+1	−1	+1
10	+1	−1	+1	−1	+1	−1	+1	+1	−1	+1	−1	+1	−1	+1	−1
11	+1	−1	+1	+1	−1	+1	−1	−1	+1	−1	+1	+1	−1	+1	−1
12	+1	−1	+1	+1	−1	+1	−1	+1	−1	+1	−1	−1	+1	−1	+1
13	+1	+1	−1	−1	+1	+1	−1	−1	+1	+1	−1	−1	+1	+1	−1
14	+1	+1	−1	−1	+1	+1	−1	+1	−1	−1	+1	+1	−1	−1	+1
15	+1	+1	−1	+1	−1	−1	+1	−1	+1	+1	−1	+1	−1	−1	+1
16	+1	+1	−1	+1	−1	−1	+1	+1	−1	−1	+1	−1	+1	+1	−1
	a	b	−ab	c	−ac	−bc	abc	d	−ad	−bd	abd	−cd	acd	bcd	−abcd

Strategy	# Factors	Which Column?
Full Factorial	4	1, 2, 4, 8
Resolution V	5	1, 2, 4, 8, 15
Screening	6-15	1, 2, 4, 8, 14, 7, 11, 13, 15, 12, 10, …

Table 2.7 L_{18} Design [1] for One 2-Level Factor and Seven 3-Level Factors

Run #	L_{18} Design							
	1	2	3	4	5	6	7	8
1	−1	−1	−1	−1	−1	−1	−1	−1
2	−1	−1	0	0	0	0	0	0
3	−1	−1	+1	+1	+1	+1	+1	+1
4	−1	0	−1	−1	0	0	+1	+1
5	−1	0	0	0	+1	+1	−1	−1
6	−1	0	+1	+1	−1	−1	0	0
7	−1	+1	−1	0	−1	+1	0	+1
8	−1	+1	0	+1	0	−1	+1	−1
9	−1	+1	+1	−1	+1	0	−1	0
10	+1	−1	−1	+1	+1	0	0	−1
11	+1	−1	0	−1	−1	+1	+1	0
12	+1	−1	+1	0	0	−1	−1	+1
13	+1	0	−1	0	+1	−1	+1	0
14	+1	0	0	+1	−1	0	−1	+1
15	+1	0	+1	−1	0	+1	0	−1
16	+1	+1	−1	+1	0	+1	−1	0
17	+1	+1	0	−1	+1	−1	0	+1
18	+1	+1	+1	0	−1	0	+1	−1

Strategy	# Factors	Which Column?
Screening	8 (max)	2-level factor in column 1
		3-level factors in columns 2-8

Table 2.8 L_{27} Design [1] for 3-Level Factors

L_{27} Design

Run #	1	2	3	4	5	6	7	8	9	10	11	12	13
1	−1	−1	−1	−1	−1	−1	−1	−1	−1	−1	−1	−1	−1
2	−1	−1	−1	−1	0	0	0	0	0	0	0	0	0
3	−1	−1	−1	−1	+1	+1	+1	+1	+1	+1	+1	+1	+1
4	−1	0	0	0	−1	−1	−1	0	0	0	+1	+1	+1
5	−1	0	0	0	0	0	0	+1	+1	+1	−1	−1	−1
6	−1	0	0	0	+1	+1	+1	−1	−1	−1	0	0	0
7	−1	+1	+1	+1	−1	−1	−1	+1	+1	+1	0	0	0
8	−1	+1	+1	+1	0	0	0	−1	−1	−1	+1	+1	+1
9	−1	+1	+1	+1	+1	+1	+1	0	0	0	−1	−1	−1
10	0	−1	0	+1	−1	0	+1	−1	0	+1	−1	0	+1
11	0	−1	0	+1	0	+1	−1	0	+1	−1	0	+1	−1
12	0	−1	0	+1	+1	−1	0	+1	−1	0	+1	−1	0
13	0	0	+1	−1	−1	0	+1	0	+1	−1	+1	−1	0
14	0	0	+1	−1	0	+1	−1	+1	−1	0	−1	0	+1
15	0	0	+1	−1	+1	−1	0	−1	0	+1	0	+1	−1
16	0	+1	−1	0	−1	0	+1	+1	−1	0	0	+1	−1
17	0	+1	−1	0	0	+1	−1	−1	0	+1	+1	−1	0
18	0	+1	−1	0	+1	−1	0	0	+1	−1	−1	0	+1
19	+1	−1	+1	0	−1	+1	0	−1	+1	0	−1	+1	0
20	+1	−1	+1	0	0	−1	+1	0	−1	+1	0	−1	+1
21	+1	−1	+1	0	+1	0	−1	+1	0	−1	+1	0	−1
22	+1	0	−1	+1	−1	+1	0	0	−1	+1	+1	0	−1
23	+1	0	−1	+1	0	−1	+1	+1	0	−1	−1	+1	0
24	+1	0	−1	+1	+1	0	−1	−1	+1	0	0	−1	+1
25	+1	+1	0	−1	−1	+1	0	0	0	−1	0	−1	+1
26	+1	+1	0	−1	0	−1	+1	+1	+1	0	+1	0	−1
27	+1	+1	0	−1	+1	0	−1	−1	−1	+1	−1	+1	0
	a	b	−ab	ab^2	c	−ac	ac^2	−bc	abc	ab^2c^2	bc^2	$-ab^2c$	$-abc^2$

Strategy	# Factors	Which Columns?
Full Factorial	3	1, 2, 5
Screening	8-13	1, 2, 5, ...

2.2.4 Other Designs

There are at least seven other types of designs that have been used by experimenters:

	References
Plackett-Burman (P-B)	1, 4
Latin Squares (Latin)	1
Hadamard Matrices	1, 2
Foldover Designs	1
Box-Behnken (B-B)	1, 2
Central Composite Designs	1, 2, 3, 4
D-Optimal Designs	1, 2, 4

Summarizing the above designs by objective based on common uses:

Levels	OBJECTIVE			
	Troubleshooting	Screen	Model	Robust
2	Hadamard D-Optimal	Hadamard/Foldover Latin P-B D-Optimal	D-Optimal	Latin P-B D-Optimal
3	D-Optimal	D-Optimal	Central Composite D-Optimal B-B	Central Composite D-Optimal B-B
More than 3 levels	D-Optimal	D-Optimal	Central Composite D-Optimal	Central Composite D-Optimal

2-18 Straight Talk on Designing Experiments

Although this may seem confusing to you, it's really more of a history lesson than anything else. A particular design was used until something more appropriate was discovered. The following approximate timeline should put all these designs in perspective:

2-Level designs	Time	3-Level designs
Full & Fractional ▶ Factorials (Fisher & Yates) # of runs = 2^K	1920's	◀ Full & Fractional Factorials (Fisher & Yates) # of runs = 3^k
	1930's	
Plackett-Burman ▶ # of runs = 4k	1946	◀ Plackett-Burman
	1957	◀ Central Composites (Box-Wilson)
	1960	◀ Box-Behnken
	1970's	
D-Optimal ▶ any # of runs	1985	◀ D-Optimal
	1990's	

We will postpone discussion of the designs used for modeling and robust designs until later in the text. The only other design of practical importance in screening and troubleshooting is the D-optimal design. Because it is the "latest and (possibly) the greatest," we believe a brief discussion is appropriate. These designs fall into a class of designs that may not be orthogonal, but nearly orthogonal.

2.3 Nearly Orthogonal Designs – the D-Optimal

Thanks to the availability of low-cost computing power and powerful software, computer-aided experimental design is rapidly gaining in popularity. One family of computer generated designs are called D-optimal designs. Benefits associated with D-optimal designs include:

1. Being able to handle any of our four objectives (screening, troubleshooting, modeling, and robustness).

2. Accommodating any number of factors or levels. For example, suppose you wished to evaluate four factors each at two levels and one factor (buffer type – which is categorical) at five levels. None of the tabled designs fit this problem well. It would be possible, however, to fit a D-optimal design to this problem in as few as nine runs.

3. Accepting both qualitative and quantitative factors (central composite designs require continuously adjustable factors).

4. Accepting any arrangement of interactions.

5. Generally require fewer runs than the appropriate orthogonal design; the minimum run D-optimal design will never require more runs.

With all the above advantages, there are, unfortunately, some disadvantages:

1. There are no simple tables of available designs.

2. These designs are computer generated and not all experimental design software packages support these designs.

3. Analysis must be done with regression, thus compounding the problem of interpreting the results.

4. D-optimal designs are not always orthogonal.

5. Design requires knowing model desired.

The minimum number of runs required for a D-optimal design can be determined as follows:

1. Determine the number of degrees of freedom (df) associated with each factor, then sum this total. The df of each factor is the number of levels minus one.

2. For each interaction required, calculate the df. Sum the total df for all interactions. For an interaction (say of factors A & B), the df = (df of factor A) × (df of factor B).

3. Add one to the sum of 1 and 2. This is the minimum number of runs required for the design.

EXAMPLE 2.1 Suppose we have the following arrangement of factors and levels:

Factor	Levels
Material type	A, B, C, D, E
Rpm	2000, 3000
Distance	1, 2
Pressure	1, 2

Additionally, we suspect a large interaction between rpm and pressure. Using the rules as shown above, the number or runs would be:

1. $df_{material} = 4$, $df_{rpm} = 1$, $df_{distance} = 1$, $df_{pressure} = 1$ Total = 7

2. $df_{rpm \times pressure} = (df_{rpm})(df_{pressure}) = (1)(1) = 1$

3. Add one; the number of runs would then be 9.

For more information regarding the origin and theory of D-optimal designs, we suggest [2],[3], and [4]. If you're only interested in an intuitive feel for the inner workings of the software, let's return to our example on page 2-1. We were interested in a designed experiment with three, two-level factors and one response. Assume interactions are not important. If we can afford an experiment with eight runs, the D-optimal algorithm will select an eight run orthogonal design. RS Discover, a BBN Software Products DOE package, produced the following eight run design (1 is the low setting and 2 is the high setting):

Run	Pull Back	Hook	Peg
1	2	1	2
2	1	1	1
3	2	2	1
4	1	2	2
5	2	1	2
6	1	1	1
7	2	2	1
8	1	2	2

For three factors at two levels, the D-optimal has given us Taguchi's L_4 with two replicates for each run. (We will use the term repetition when we intend to complete a run more than once without resetting the factors. We will refer to replication when we intend to complete a run more than once, resetting the factors each time. Often we will use the term "replicates" to enumerate the number of data points collected per run. We will discuss this further in Chapter 3.) Using the tree diagram from page 2-2 to show the paths selected by the D-optimal design, you can assemble the L_4:

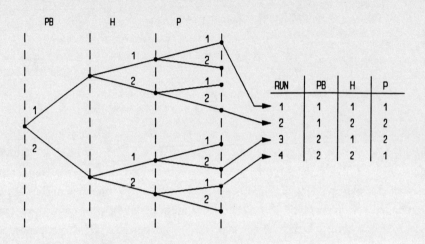

You should have also noticed that the algorithm selected two paths from the lower and upper halves of the tree.

NOTE: The four paths not selected in the above design also form an orthogonal array. This is the matrix commonly attributed to classical experimental designers:

Run	Pull Back	Hook	Peg
1	1	1	2
2	1	2	1
3	2	1	1
4	2	2	2

⇔

Run	Pull Back	Hook	Peg
1	−	−	+
2	−	+	−
3	+	−	−
4	+	+	+

On the other hand, if we could only afford seven runs (with no important interactions), the design would not be orthogonal. Again, using RS Discover we generated the following seven run design:

Selecting an Orthogonal Array 2-23

Run	PB	H	P
1	1	1	2
2	2	1	1
3	1	2	1
4	2	2	2
5	1	1	1
6	2	2	1
7	2	1	2

Relating this selection of runs to the tree diagram on the previous page, the software chose three runs from the top half of the tree and four runs from the bottom half.

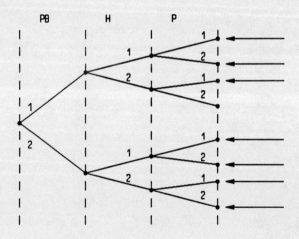

Of course, this design is not perfectly orthogonal. This type of design should be used when time and money constraints prevent the use of an orthogonal design. For another example of a D-optimal design, see Case Study II-2 at the end of Section II.

2.4 Orthogonal vs What???

Although we have just suggested using a "nearly" orthogonal design (D-optimal), there are advantages to using an orthogonal design. The major advantages of full factorial and/or orthogonal arrays are:

1. We can develop a predictive model for responses in the range of interest.
2. We can estimate interactions.
3. Our analysis will be greatly simplified.
4. Factors and interactions are independent.

Having said that, the two most common, BUT NOT RECOMMENDED, methods of experimenting in today's industries are still:

1. Best guess.
2. One-factor-at-a-time.

Best guess could obviously give a good guesser a very good answer, without any of the above advantages (although it is probably rather simple). The one-factor-at-a-time approach is simply stepping through an experiment by changing "one factor at a time." The following example illustrates this approach:

EXAMPLE 2.2 Consider the same three statapult factors as above:
    Pull Back
    Hook
    Peg

Each can be tested at two levels. The one-factor-at-a-time approach prescribes that we begin by measuring in-flight distance with all factors set at a certain level:

Pull Back	Hook	Peg	Avg Distance
−	−	−	72.33

Then we would measure it again with only one factor changed.

Pull Back	Hook	Peg	Avg Distance
−	−	−	72.33
+	−	−	35.67

If we were searching for a maximum downrange distance, we would leave the pull back angle at the setting that gives us the greatest distance for the remainder of the test. The next change would be to the hook setting to see if we can improve the downrange distance.

Pull Back	Hook	Peg	Avg Distance
−	−	−	72.33
+	−	−	35.67
−	+	−	138.5

We would stick with the hook setting that yields greatest distance and change the setting for the peg. Finally we would set the peg at the level that achieves maximum distance.

Pull Back	Hook	Peg	Avg Distance
−	−	−	72.33
+	−	−	35.67
−	+	−	138.5
−	+	+	209.0

With a little luck, we can pick the winner. Obviously we have only tested half of the possible combinations; we do not have an orthogonal design and therefore cannot attribute any effects to a particular factor; we cannot distinguish the effect of a factor from any interaction; and we have no idea if this is really an optimum. Fortunately, we will be able to confirm (or refute) this result in a couple of chapters.

2.5 Summary

The following two charts summarize the advantages and disadvantages of the different designs for two and three levels:

2.5.1 Two-Level Designs
(Factor screening/estimate first order linear models)

The designs we have discussed up to this point are:

Type	Why?	Advantages	Disadvantages
Full factorial	Modeling Troubleshooting	Test all possibilities	Cost Time
Fractional factorial	Screening Troubleshooting	Estimate linear effects with few runs; Estimate some/no interactions	Some aliasing
D-optimal	All: Screening Troubleshooting Modeling Robust Design	Any # of levels OK; Mixed levels OK; Estimate all, some, or no interactions; Minimum # of runs	SW required New More complex Not always orthogonal

The designs we will defer to our next text are:

Type	Why?	Advantages	Disadvantages
Foldover (Not used much in applications)	Sequential screening; Expensive experimentation	$R_{III} \rightarrow R_{IV}$ Learn as you go	Technology must support.
Plackett-Burman (Multiples of 4)	Screening Troubleshooting (Taguchi uses for robust design)	Lots of factors in few runs	Confounding bad for modeling

2.5.2 Three-Level Designs

The designs we will discuss are:

Type	Why?	Advantages	Disadvantages
Full factorial	Not used often	Estimate linear effects. Estimate quadratic effects. Estimate all possible interactions.	Expensive Resources
Box-Wilson (CCD) Includes CCC and CCF	Modeling Robust Designs	Very efficient 2nd order tool. Estimates linear & quadratic effects. Estimate all, some, or no interactions. May be rotatable. Accommodates sequential testing.	Need continuous factors. May be slightly nonorthogonal. Best suited for quantitative factors.
D-Optimal	Screening Troubleshooting Modeling Robust Designs	Same as in previous table.	Same as in previous table.

The designs we will defer to our next text are:

Type	Why?	Advantages	Disadvantages
Latin Squares	Screening Troubleshooting Robust Designs	Lots of factors in few runs. Estimate linear & quadratic effects. Categorical or continuous factors.	SEVERE confounding.
Box Behnken	Modeling	Resolution V Estimate main & quadratic effects. Estimate all 2-way interactions.	Slightly nonorthogonal. # of runs large. Estimates all 2-way interactions & 2nd order effects, whether they are needed or not.

Chapter 2 Bibliography

1. S. R. Schmidt and R. G. Launsby, *Understanding Industrial Designed Experiments (3rd edition).* Colorado Springs, CO: Air Academy Press, 1991.

2. Box and Draper, *Empirical Model-Building and Response Surfaces.* Wiley.

3. J. A. Cornell, *Experiments with Mixtures.* Wiley.

4. RS Discover Users Guide, BBN Software Products Corporation, Cambridge MA.

CHAPTER 3
CONDUCTING YOUR EXPERIMENT

Excerpt from 2-1

   III.   CONDUCT THE EXPERIMENT AND RECORD THE DATA

Now that we have chosen an orthogonal array, what's next? Essentially, it is time to take your design matrix, run your experiment, and collect some data. Although this is sometimes easier said than done, this chapter will get you started and give you some ideas to ponder, as well as some common errors to avoid.

3.1 The Worksheet

Your first step is to make a copy of the orthogonal array — your design matrix. This will be your worksheet. For our statapult experiment we will use an L_8 (Table 2.3). As you recall, the settings for each factor in the orthogonal array are ± 1. In your experiment, these orthogonal settings may be confusing. We recommend you go through the table and change the settings for the main factors to their "real" settings as shown in Table 3.1. If you are working from tables this will be more than sufficient. If you are working with a software package, most have the ability to print the matrix in a worksheet format. This will be simply the design matrix with the appropriate number of places to record your measured responses for each run. Remember, each run is the experimenter's "recipe" for setting the control factors. For computer generated worksheets, the real values will already be substituted for the orthogonal values. Using a simple package such

Table 3.1 L_8 with Real Settings

Run #	1	2	3	4	5	6	7
				L_8 Design			
1	−1 490	−1 1	−1	−1 1	−1	−1	−1
2	−1 490	−1 1	−1	+1 4	+1	+1	+1
3	−1 490	+1 5	+1	−1 1	−1	+1	+1
4	−1 490	+1 5	+1	+1 4	+1	−1	−1
5	+1 440	−1 1	+1	−1 1	+1	−1	+1
6	+1 440	−1 1	+1	+1 4	−1	+1	−1
7	+1 440	+1 5	−1	−1 1	+1	+1	−1
8	+1 440	+1 5	−1	+1 4	−1	−1	+1
	a (Pullback Angle)	b (Hook)	−ab	c (Peg)	−ac	−bc	abc

as Q-Edge [1], the worksheet for the statapult looks like:

Run	A	B	*	C	*	*	*
1	490	1		1			
2	490	1		4			
3	490	5		1			
4	490	5		4			
5	440	1		1			
6	440	1		4			
7	440	5		1			
8	440	5		4			

Factor A is the pull back angle, factor B the hook position, and factor C the peg position. No matter which software package you use or whether you simply copy one of the tabled designs, you need to give some thought to the three R's:

- Randomization
- Repetition
- Replication

3.1.1 Randomization

This is a decision to order your runs randomly. The idea centers around variation in your process. If you would like to spread out variation that may be induced by run set-up, different operators, etc., or if you'd just like to counter unknown variation, this is the way to go. However, this is also a question of time and money. Do you have the resources to possibly change all of your factor settings after every run?

If not, there are alternatives. Whether you decide to randomize your runs or not, it will be up to you to assign factors to particular columns. The matrix in Table 3.1 has four -1's followed by four $+1$'s in the first column. If you have a factor that is very difficult to change, and you intend to use the array as ordered in the table, then you would probably assign that factor to that column. A possible candidate could be the temperature in a chemical bath. It is easy to raise the temperature, but difficult (or time consuming) to cool it down. Some software packages handle this problem by having you rate the difficulty of changing a factor setting. RS Discover uses "easy, moderate, and hard." That way, the computer can assign a factor that is difficult to change to the "first column" as we did above. It may also be able to randomize other factor settings within the low/high "blocks" of the "difficult to change" factor. It is up to you to determine the capabilities of your software package. We recommend that you randomize your runs if it fits into your budget.

3.1.2 Repetition vs Replication

In addition to randomization, we must decide on how many responses we would like to record for each run, and how we plan to order them. If we are trying to reduce variation, we will need more samples than if we are trying to shift the average. You will find a good discussion of sample sizes in [2]. For the statapult we plan to collect three data points for each run. That will give us 24 "cells" of data for the experiment.

In Chapter 2 we said that we would refer to repetition when we intend to complete a run more than once without resetting the factors. A replication occurs when we intend to complete a run more than once, resetting the factors each time. Replication will tend to provide a better estimate of experimental error, but it may be expensive to reset all factors for each run. During our seminar we have our students do their experiments both ways. The Q-Edge worksheet leaves the three R's up to you, giving you a column for each repeated response. Our statapult worksheet with three data points per run looks like the following:

3-4 Straight Talk on Designing Experiments

Run	A	B	*	C	*	*	*	D1	D2	D3
1	490	1		1				①	②	③
2	490	1		4						
3	490	5		1						
4	490	5		4				⑤		
5	440	1		1				⑧		
6	440	1		4				⑥		
7	440	5		1				④	⑨	
8	440	5		4				⑦		

If pull back is set at 490, hook at 1, and peg at 1 (run 1) and three in-flight distances are measured, the data will be placed in the positions marked ①, ②, and ③. That is an example of repetition. If the factors are set at the levels specified in the seventh row and one in-flight distance is measured, the data will be stored in the position marked ④. The settings are then changed and other rows are run. Eventually the factors will be reset to the row 7 levels. The response data will be placed in the position marked ⑨. The data stored in positions ④ and ⑨ is replicated data. Other, more sophisticated software will give you this same worksheet as 24 rows, if you decide to run replications. Again, most experimenters are limited by their resource constraints to repetitions.

3.2 Common Errors

Before you run to the lab or the line with your worksheet in hand, we should discuss some common experimental errors. Some of these may make more sense after we analyze our data in Chapter 4, but if we touch on them now we may save some of that time and money we've talked about. Although most texts spend a lot of time explaining orthogonal arrays and analysis, most mistakes and problems will occur during planning and conducting an experiment, or by failing to confirm the final results. After helping with hundreds of experiments, we believe the causes for poor results can be lumped into the following seven reasons:

   a. Lack of experimental discipline
   b. Measurement error
   c. Too much variation in the response

d. Aliased effects
e. Inadequate model
f. Something changed
g. Improper experimental region

3.2.1 Lack of Experimental Discipline

If your design specifies low pull back angle, low hook, and low peg settings for the first experimental run, that is exactly the way it should be run. However, operators/technicians are often given the flexibility to change settings to prevent a bad part from being manufactured. And it should be no surprise that low levels of the three main factors in our statapult example should yield a low downrange distance. If you know it, so does the operator who does this every day. The operator may want to prevent an obviously low downrange distance and may be tempted to alter your settings — thinking he/she is helping you out, not understanding your need for a balanced design. We label that as a lack of experimental discipline. We can combat this error with a detailed set of operator instructions (see Section 3.3), by simply monitoring the experiment ourselves, or making the operators part of the team in the first place. This is particularly important if you are not actually conducting the experiment.

3.2.2 Measurement Error/Too Much Variation

If your measuring device or system is inaccurate or unstable, you will introduce additional error into your process. Example 3.1 is an experimental design performed on an unstable measuring device used to measure computer disk thickness. Measurement error may manifest itself as too much variation in your process response. Because we make an eyeball estimate of the in-flight distance of a ball fired by our statapult, the induced error can be rather large. We encourage our seminar students to think of ingenious ways to pinpoint the impact point.

3.2.3 Aliased Effect/Inadequate Model

Due to lack of knowledge about your process or a poor design, you may have aliased the wrong factors or interactions, or made an incorrect assumption of linearity. In either case, the predictive model we develop may not confirm our responses.

3.2.4 Something Changed

If you obtain strange results, do some digging. Someone may have started using a new "lot" of material midway through the experiment; a new shift may have punched in between runs; the morning sunshine may have been replaced with an afternoon downpour raising the indoor humidity; etc. If you can control these factors, you may want to consider some of them as noise factors and employ a robust design (Chapter 7).

3.2.5 Improper Experimental Region

The last reason for poor results is that we cannot find an optimum because we are not experimenting within the right ranges of our factors. Hopefully, our analysis will point us in the right direction. This is also a good time to **caution** against extrapolating outside the domain of experimentation. A linear relationship within our domain of interest does not guarantee a linear relationship outside that region.

3.3 Operator Instructions

Although most of the reasons for poor results are out of the control of the operator (or experimenter), some are not. We will try to prevent those with a set of specific operator instructions. If you intend to conduct your own experiment this may not be as important as if you are going to hand it off to an "operator." In either case, a set of instructions will force the designer to think through the experiment and give the operator some guidance.

There are no real rules for these instructions, but you might consider including some of the following:

1. Background
2. Experimental objective
3. Response(s)
4. Factors
5. Factor levels
6. Dress requirements
7. Equipment condition requirements
8. Equipment set-up directions
9. Specific directions for each run

10. Equipment shut-down instructions
11. Who to call when...with phone numbers!
12. Data sheets to record:
    a. Date/Time
    b. Temperature
    c. Operator name/number
    d. Actual equipment set-up
    e. Equipment serial numbers
    f. Material lot
    g. Responses for each run
    h. Actual factor settings for each run

These are not all-inclusive, but should get you started. Since everyone likes examples, these next three pages are excerpts from a set of operator instructions at a disk manufacturing company. The first page simply sets the stage for you, while the second and third pages were given to each operator before each run.

EXAMPLE 3.1 Instructions to the operator for a designed experiment on a computer disk were developed to combat some of the reasons for experimental error listed in Section 3.2.

INTRODUCTION: The beta-backscatter technique for measuring nickel thickness on aluminum substrates is known to have wide variations. Previous correlation between two company divisions confirms the problem. The purpose of this experiment is to further optimize the measuring technique to reduce the variations.

EXPERIMENTAL OBJECTIVE: To determine the effects of different variables on the repeatability of the NI thickness measurement and to reduce the variation between readings.

IDENTIFICATION OF DEPENDENT VARIABLES (RESPONSES):
Thickness measurement

IDENTIFICATION OF INDEPENDENT VARIABLES (FACTORS):
CONTROL VARIABLES:
(A) APERTURE TYPE
 LEVEL 1 = ALUMINUM SIZE 1 UNCOATED
 LEVEL 2 = STAINLESS STEEL SIZE 2 UNCOATED
 LEVEL 3 = ALUMINUM SIZE 3 UNCOATED
 LEVEL 4 = ALUMINUM SIZE 2 COATED
 LEVEL 5 = ALUMINUM SIZE 2 UNCOATED
 LEVEL 6 = ALUMINUM SIZE 1 COATED

(E) READING TIME
 LEVEL 1 = TIME 1
 LEVEL 2 = TIME 2
 LEVEL 3 = TIME 3

(H) NUMBER OF READINGS
 LEVEL 1 = 2
 LEVEL 2 = 4
 LEVEL 3 = 6

(K) FIXTURE TYPE
 LEVEL 1 = A
 LEVEL 2 = B
 LEVEL 3 = MANUAL

(L) UNIT NUMBER
 LEVEL 1 = 1
 LEVEL 2 = 2

Based on the control factors and levels, what orthogonal design do you recommend? The designers had five qualitative factors: one at 6 levels, two at 3 levels, and two at 2 levels. They had only one continuous factor (TIME) at 3 levels. The experimenters were interested in screening out the insignificant main factors, so interactions were not considered important. Hopefully, you are ready to recommend a D-optimal design. Take a few minutes to come up with the minimum number of runs required for this design. We think you will need 14 runs. Continuing with the example, here is page 2:

DISK SERIAL # _____
     RUN # _____

DATE _____   ROOM TEMP. _____   OPERATOR _____

OPERATOR DRESS REQUIREMENTS
   The operator shall wear a facemask, bouffant hair cover, frock and gloves at all times while performing this experiment.

EQUIPMENT CLEANING REQUIREMENTS
   At the beginning of each run, thoroughly wipe the equipment and equipment tables with a dry Techni-Cloth. Using an air gun, spray all NI equipment and the surrounding area to ensure that it is free of dust.

UNIT SET-UP

   Warm up time (Minimum of 30 minutes) _____
   Unit # _____
   Power Line Filter _____
   Probe Fixture Type _____
   Probe Serial # _____
   Aperture _____

CALIBRATION
   Perform calibration per NI Calibration Procedure making sure to calibrate on the marked area of the standard.

   Reading Time _____
   # of Readings _____

In this particular example, the experimenters wanted to collect a lot of data. As you can see on the page below, they took 36 readings. They averaged every six data points and called each of those averages a measurement. They later entered those six measurements into the response columns of their design matrix. Repetitions were used instead of replications. Note that the operators were not given the complete design matrix. They were given the required settings for each run on the preceding page and collected data, for that run only, on the page below. A new set of instructions and a data collection form were passed out prior to each run.

The last page of the instructions was as follows:

RUN #_____

DATA COLLECTION

(A) Before loading the disk into the NI fixture, carefully clean it using an air gun.
(B) Load the disk into the hub with the marked area up and at 12:00.
(C) Lower the probe gently so it is touching the disk in the marked area.
(D) Press the start button on the NI Unit and record the $\mu"$ reading.
(E) Repeat step D until a set of readings has been taken. The average of this set of readings = one measurement.
(F) Raise the probe and clean the disk and probe tip using the air gun.
(G) Repeat steps C through F until six sets of readings (6 measurements) have been taken.

READINGS   MEASUREMENT

___ ___ ___ ___ ___ ___   ___
___ ___ ___ ___ ___ ___   ___
___ ___ ___ ___ ___ ___   ___
___ ___ ___ ___ ___ ___   ___
___ ___ ___ ___ ___ ___   ___
___ ___ ___ ___ ___ ___   ___

SHUT DOWN PROCEDURE

At the end of each session, set up the production NI Unit #1 so it is ready for production. This includes calibration with the production probe and running the reference standard. Cover NI Unit #2.

Your process will probably differ extensively from this disk manufacturer's. Try to account for as many contingencies as you can before the experiment. Remember, this is not an attempt to formalize the experimental procedure. This is an approach that may save your experiment!

Chapter 3 Bibliography

1. *Q-Edge Software*. Available from Launsby Consulting, Colorado Springs, CO.

2. S. R. Schmidt and R. G. Launsby, *Understanding Industrial Designed Experiments (3rd edition)*. Colorado Springs, CO: Air Academy Press, 1991.

CHAPTER 4
ANALYZING YOUR RESULTS

> XI. ANALYZE THE DATA, DRAW CONCLUSIONS, MAKE PREDICTIONS, AND DO CONFIRMATORY TESTS.

Now that we have massive amounts of data, we have the (sometimes) formidable task of sifting through it. With some simple analysis techniques we will be able to sort out what is important and what is not. We will rely on averages and graphs to give us most of our insight — particularly for designs involving two levels. If analyzing averages is not enough, we will estimate and analyze population variances from our sample to give us an additional measure of significance of factors. Once we try to analyze three-level designs, we are forced to use regression to do our analysis. Each of these is referred to by the following names:

ANOM - ANalysis Of Means (Examining averages)
ANOVA - ANalysis Of VAriance (Examining variances)
Multiple Regression - Using the method of least squares to develop predictions.

We will also use the following graphs when they apply:

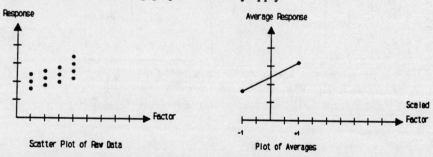

Figure 4.1a Commonly Used Graphics

4-2 Straight Talk on Designing Experiments

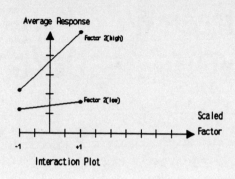
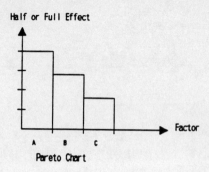

Figure 4.1b Commonly Used Graphics

The best way to illustrate this is probably through an example of a two-level design. Let's continue with the statapult example we began in Chapter 1 using the L_8 orthogonal array we introduced in Chapter 2. With the worksheet from page 3-4 in hand, we ran the experiment and collected the following data:

Table 4.1 Completed Worksheet for the Statapult Experiment

Run	A	B	*	C	*	*	*	D1	D2	D3
1	490	1		1				85	74	58
2	490	1		4				128	132	122
3	490	5		1				150	145	120.5
4	490	5		4				215	208	204
5	440	1		1				32.5	36	38.5
6	440	1		4				69	80	68
7	440	5		1				68	81.5	68.5
8	440	5		4				114	101.5	110.5

Note: The responses D1, D2, and D3 could be the responses at three different noise levels. From our example, this could be distances when air is calm (D1), mild breeze (D2), and in a stiff breeze (D3).

Since we will be trying to build a mathematical equation that predicts distance, we will use the letter y for the response. Using "y" will keep us consistent with most other texts.

4.1 Analysis of Means and Graphical Analysis

Step 1: Calculate $\bar{y}$ (average y for each run, $\bar{y} = \dfrac{\sum y_i}{n_r}$). Put matrix in coded form.

Table 4.2 Worksheet with Means Completed

Run	A	B	−AB	C	−AC	−BC	ABC	y_1	y_2	y_3	$\bar{y}$
1	−	−	−	−	−	−	−	85	74	58	72.33
2	−	−	−	+	+	+	+	128	132	121	127
3	−	+	+	−	−	+	+	150	145	120.5	138.5
4	−	+	+	+	+	−	−	215	208	204	209
5	+	−	+	−	+	−	+	32.5	36	38.5	35.66
6	+	−	+	+	−	+	−	69	80	68	72.33
7	+	+	−	−	+	+	−	68	81.5	68.5	72.66
8	+	+	−	+	−	−	+	114	101.5	110.5	108.66

Step 2: Calculate s (standard deviation of the responses in any run) and ln s

$$ s = \sqrt{\dfrac{\sum (y_i - \bar{y})^2}{n - 1}} $$

Table 4.3 Worksheet with s and ln s Computed

Run	A	B	−AB	C	−AC	−BC	ABC	y_1	y_2	y_3	$\bar{y}$	s	ln s
1	−	−	−	−	−	−	−	85	74	58	72.33	13.577	2.608
2	−	−	−	+	+	+	+	128	132	121	127	5.568	1.717
3	−	+	+	−	−	+	+	150	145	120.5	138.5	15.788	2.759
4	−	+	+	+	+	−	−	215	208	204	209	5.568	1.717
5	+	−	+	−	+	−	+	32.5	36	38.5	35.66	3.014	1.103
6	+	−	+	+	−	+	−	69	80	68	72.33	6.658	1.896
7	+	+	−	−	+	+	−	68	81.5	68.5	72.66	7.654	2.035
8	+	+	−	+	−	−	+	114	101.5	110.5	108.66	6.449	1.864

Step 3: Calculate the average of the averages $\left(\bar{\bar{y}}\right)$, the average of s $(\bar{s})$, and the average of the ln s $(\overline{\ln s})$.

$$\bar{\bar{y}} = 104.521$$
$$\bar{s} = 8.035$$
$$\overline{\ln s} = 1.9624$$

NOTE: We recommend ln s over a signal/noise ratio based on [1], ease of use, and interpretation.

Step 4: Eyeball Check!

The following scatterplots of raw data may give you some useful information. If nothing else, this will be your first cross check of the data.

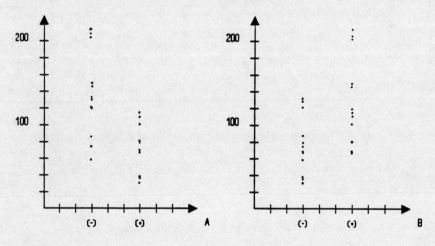

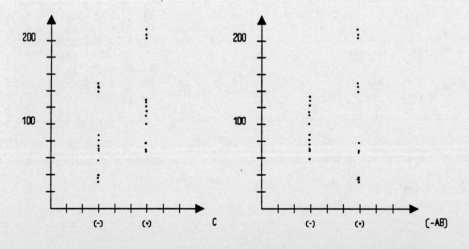

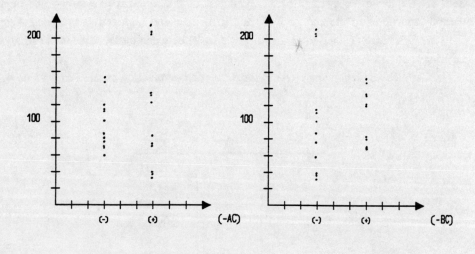
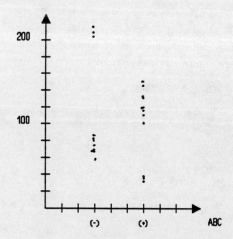

Figure 4.2 Scatterplots of Raw Data for Each Factor

Clearly, some of the factors have an effect on the downrange distance. They also have the sort of effect we might expect — as the arm is pulled further back, or the hook or peg is adjusted to increase the tension in the rubber band, the projectile goes further. That matches our technical knowledge of the process.

Step 5: Calculate the average of the averages, of the standard deviations, and of the natural logarithms of the standard deviations for low and high settings of each factor. Find the difference (Δ) and the 1/2 difference (Δ/2) between those. This is the Analysis Of Means.

Let's use pullback angle (A) to illustrate the computation of Δ. From Table 4.3:

Run	A	$\bar{y}$
1	−	72.33
2	−	127.00
3	−	138.50
4	−	209.00
5	+	35.67
6	+	72.33
7	+	72.67
8	+	108.67

The average response when A is low is:

$$\bar{y}_- = \frac{72.33 + 127.00 + 138.50 + 209.00}{4} = 136.71$$

The average response when A is high is:

$$\bar{y}_+ = \frac{35.67 + 72.33 + 72.67 + 108.67}{4} = 72.33$$

The effect (Δ) of factor A on the response is $\bar{y}_+ - \bar{y}_-$.

$$\Delta = \bar{y}_+ - \bar{y}_- = 72.33 - 136.71 = -64.37$$

The half effect (Δ/2) is half of the effect. The process is repeated for the remaining average responses, as well as s and ln s.

Table 4.4 Analysis of Means

Run	A	B	−AB	C	−AC	−BC	ABC	ȳ	s	ln s
1	−	−	−	−	−	−	−	72.33	13.58	2.61
2	−	−	−	+	+	+	+	127.00	5.57	1.72
3	−	+	+	−	−	+	+	138.50	15.79	2.76
4	−	+	+	+	+	−	−	209.00	5.57	1.72
5	+	−	+	−	+	−	+	35.66	3.01	1.10
6	+	−	+	+	−	+	−	72.33	6.66	1.90
7	+	+	−	−	+	+	−	72.66	7.65	2.04
8	+	+	−	+	−	−	+	108.66	6.45	1.86

		A	B	−AB	C	−AC	−BC	ABC
ȳ	Avg−	136.71	76.83	95.17	79.79	97.96	106.42	106.58
	Avg+	72.33	132.21	113.88	129.25	111.08	102.63	102.46
	Δ	−64.38	55.38	18.70	49.46	13.12	−3.79	−4.12
	Δ/2	−32.19	27.69	9.35	24.73	6.56	−1.90	−2.06
s	Avg−	10.13	7.20	8.31	10.01	10.62	7.15	8.36
	Avg+	5.94	8.86	7.76	6.06	5.45	8.92	7.70
	Δ	−4.19	1.66	−.55	−3.95	−5.17	1.77	−.66
	Δ/2	−2.10	.83	−.28	−1.98	−2.59	.89	−.33
ln s	Avg−	2.20	1.83	2.06	2.13	2.28	1.82	2.06
	Avg+	1.73	2.09	1.87	1.80	1.64	2.10	1.86
	Δ	−.47	.26	−.19	−.33	−.64	.28	−.20
	Δ/2	−.24	.13	−.09	−.17	−.32	.14	−.10

Step 6: Plot the averages (or marginals). To build a plot of averages we simply plot the average high and the average low for each factor. Then, connect the dots (per factor). For factor A,

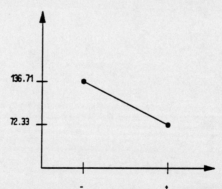

Figure 4.3a Plot of Average y for Pull Back Angle

For all factors,

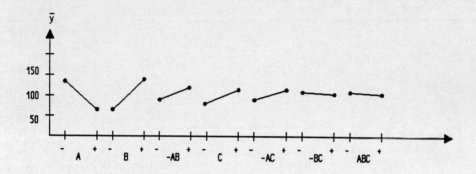

Figure 4.3b Plot of Average y

The sign of the slope should match the sign of Δ for that factor. The greater (+ or −) the slope, the more significant the effect. It should be obvious that factors A, B, and C most significantly effect the average response.

Doing the same for s, you see that A, C, and −AC seem to have the most effect on variation.

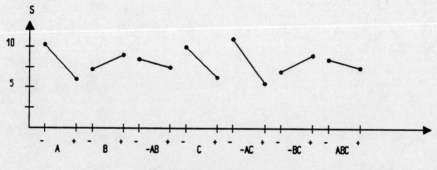

Figure 4.4 Plot of Average s

For ln (log$_e$) s we see the same trends as shown for s, lending support to our decision to use ln s as our measure of variation.

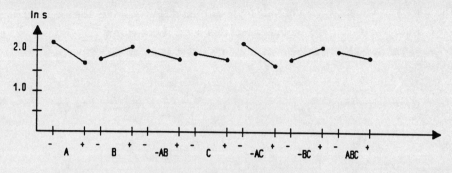

Figure 4.5 Plot of Average ln s

Step 7: To demonstrate whether there's an interaction between factors, we could build an interaction plot. The construction of an interaction plot can be tricky. To build the AB interaction plot for the average responses, we will need to use Table 4.4:

Run	A	B	$\bar{y}$
1	−	−	72.33
2	−	−	127.00
3	−	+	138.50
4	−	+	209.00
5	+	−	35.67
6	+	−	72.33
7	+	+	72.67
8	+	+	108.67

You may place either factor on the horizontal axis:

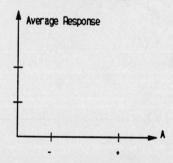

Figure 4.6a Interaction Plot Set Up

The lines you are about to plot will represent the different levels for the other factor. Since B has two levels, you should expect two lines. For the first line (B low), we need to know the average response when A is low and A is high.

Average response for B low and A low: $\dfrac{72.33 + 127.00}{2} = 99.67$

Average response for B low and A high: $\dfrac{35.67 + 72.33}{2} = 54.00$

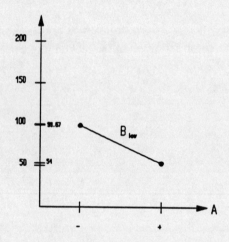

Figure 4.6b B low

For the second line (B high):

Average response for B high and A low: $\dfrac{138.50 + 209.00}{2} = 173.75$

Average response for B high and A high: $\dfrac{72.67 + 108.67}{2} = 90.67$

The final interaction plot for AB looks like:

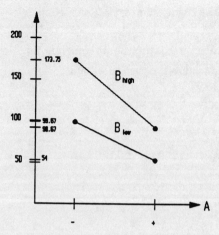

Figure 4.6c AB Interaction Plot

Now, looking at all two-factor interactions:

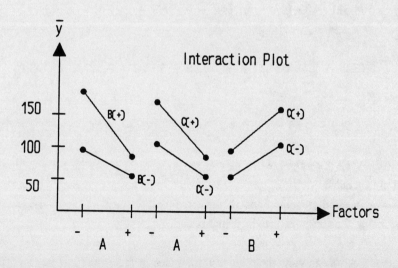

Figure 4.6d Interaction Plot for All 2-Factor Interactions

4-12 Straight Talk on Designing Experiments

If the slopes are not equal, there may be some interaction. It appears that the AB and AC interactions may be stronger than the BC interaction.

Step 8: Build the prediction equations. In general, for 2-level orthogonally coded designs, the prediction equation for each of these is:

$$\hat{y} = \bar{\bar{y}} + \frac{\Delta A}{2}A + \frac{\Delta B}{2}B + \frac{\Delta A \cdot B}{2}A \cdot B + ...$$

$$\hat{s} = \bar{s} + \frac{\Delta A}{2}A + \frac{\Delta B}{2}B + \frac{\Delta A \cdot B}{2}A \cdot B + ...$$

$$\ln \hat{s} = \overline{\ln s} + \frac{\Delta A}{2}A + \frac{\Delta B}{2}B + \frac{\Delta A \cdot B}{2}A \cdot B + ...$$

where the "^" indicates "predicted." The $\frac{\Delta \text{letter}}{2}$ is the $\frac{\Delta}{2}$ value you have just calculated for each factor and the attached letter is a variable (to be evaluated with numbers between or equal to -1 and $+1$).

In our example, our prediction equations are:

$$\hat{y} = 104.521 - 32.19A + 27.69B - 9.35AB + 24.73C - 6.56AC + 1.90BC - 2.06ABC$$

$$\hat{s} = 8.0345 - 2.09A + .83B + .28AB - 1.97C + 2.58AC - .88BC - .33ABC$$

$$\ln \hat{s} = 1.962 - 2.09 + .13B + .09AB - .10C + .32AC - .14BC - .10ABC$$

Note that we have not attempted to narrow the number of terms in our equation. For now, all are included.

If you need these equations in terms of real values, you may convert these equations using the formulas provided in Appendix A.

Step 9: Another way to view the relative importance of factor and interaction effects is to use a Pareto chart, plotting $|\Delta/2|$ (half effects) as shown in Figure 4.7.

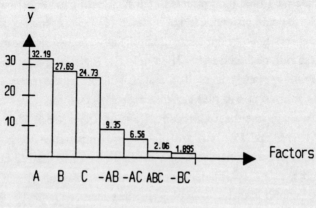
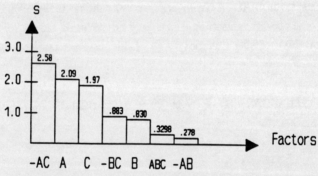
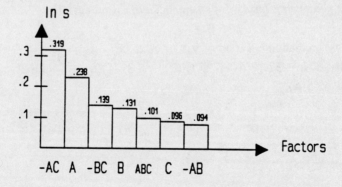

Figure 4.7 Pareto Charts for $\bar{y}$, s, and ln s

Step 10: Include only important effects in the final prediction equation.

For ln ŝ, the rule of thumb is $|\Delta/2| \geq .5$ [1]. Since none of the effects is this large, then it appears none of the factors significantly change the variability.

For ŝ, the rule of thumb is $|\Delta/2| \geq \bar{s}/2$ [1]. It also suggests none of the factors change the variability (analysis of ln s and s will tell us the same thing, hence we frequently only analyze ln s to find variance reduction factors).

For ŷ, we can use the regression output in Table 4.6. The columns headed variable, coefficient, and P(2 tail) are of practical importance to us. For a 2-level orthogonal design, the coefficient is simply the half-effect and the P(2-tail) value tells us about the significance of a particular factor/interaction. For models of the raw data, a common rule of thumb is that a P(2-tail) value less than .10 indicates the coefficient in question is likely to be different from zero and should be included in our prediction equation. All other coefficients should probably be excluded (some software packages automatically do this). Using the above rule of thumb, our prediction equation for the mean y is:

$$\hat{y} = 104.521 - 32.19(A) + 27.69(B) - 9.35(A)(B) + 24.73(C) - 6.56(A)(C)$$

If you are happy with rules of thumb and the bottom line relative to the predictive model, we are done. If you wish to obtain a more rigorous understanding of what the rest of the terms in the table mean, or how they are derived continue with Section 4.2.

Before proceeding further let's summarize our findings:

1. The best prediction equation of ŷ appears to be:
$$\hat{y} = 104.521 - 32.19(A) + 27.69(B) - 9.35(A)(B) + 24.73(C) - 6.56(A)(C)$$

2. Analysis of ln ŝ suggests there are no factors which provide reduced variance settings.

4.2 Interpreting Computer Results

If we enter our data into a software package:

Table 4.5 Format for MYSTAT Software Package Data Entry

Run	A	B	AB	C	AC	BC	ABC	y
1	−1	−1	+1	−1	+1	+1	−1	85
2	−1	−1	+1	+1	−1	−1	+1	128
3	−1	+1	−1	−1	+1	−1	+1	150
4	−1	+1	−1	+1	−1	+1	−1	215
5	+1	−1	−1	−1	−1	+1	+1	32.5
6	+1	−1	−1	+1	+1	−1	−1	69
7	+1	+1	+1	−1	−1	−1	−1	68
8	+1	+1	+1	+1	+1	+1	+1	114
9	−1	−1	+1	−1	+1	+1	−1	74
10	−1	−1	+1	+1	−1	−1	+1	132
11	−1	+1	−1	−1	+1	−1	+1	145
12	−1	+1	−1	+1	−1	+1	−1	208
13	+1	−1	−1	−1	−1	+1	+1	36
14	+1	−1	−1	+1	+1	−1	−1	80
15	+1	+1	+1	−1	−1	−1	−1	81.5
16	+1	+1	+1	+1	+1	+1	+1	101.5
17	−1	−1	+1	−1	+1	+1	−1	58
18	−1	−1	+1	+1	−1	−1	+1	121
19	−1	+1	−1	−1	+1	−1	+1	120.5
20	−1	+1	−1	+1	−1	+1	−1	204
21	+1	−1	−1	−1	−1	+1	+1	38.5
22	+1	−1	−1	+1	+1	−1	−1	68
23	+1	+1	+1	−1	−1	−1	−1	68.5
24	+1	+1	+1	+1	+1	+1	+1	110.5

and ask it to model the data linearly,

$$\text{Model } y = \text{constant} + A + B + AB + C + AC + BC + ABC$$

the result is the following multiple regression output.

Table 4.6 Multiple Regression Output Table for Statapult Data

DEP VAR: Y   N: 24   MULTIPLE R: .990   SQUARED MULTIPLE R: .979
ADJUSTED SQUARED MULTIPLE R: .970   STANDARD ERROR OF ESTIMATE: 9.008

Variable	Coefficient	Std Error	Std Coeff	Tolerance	T	P(2 Tail)
Constant	104.521	1.839	0.000		56.843	0.000
A	−32.188	1.839	−0.630	.100E+01	−17.505	0.000
B	27.688	1.839	0.542	.100E+01	15.058	0.000
AB	−9.354	1.839	−0.183	.100E+01	−5.087	0.000
C	24.729	1.839	0.484	.100E+01	13.449	0.000
AC	−6.563	1.839	−0.129	.100E+01	−3.569	0.003
BC	1.896	1.839	0.037	.100E+01	1.031	0.318
ABC	−2.063	1.839	−0.040	.100E+01	−1.122	0.279

ANALYSIS OF VARIANCE

SOURCE	SUM-OF-SQUARES	DF	MEAN-SQUARE	F-RATIO	P
Regression	61261.906	7	8751.701	107.852	0.000
Residual	1298.333	16	81.146		

4.2.1 Familiar Items

There is obviously a lot of mysterious stuff here, but there are also some familiar numbers. We see that the constant in $\hat{y}$ is 104.5, and the coefficients of A, B, AB, C, AC, BC, and ABC are as shown in column 2 of Table 4.6, respectively. That at least confirms our previous calculations.

The question that the rest of this output answers is how well does the prediction equation, model our response in the range of interest. This "goodness of fit" will be evaluated as a whole and in parts.

4.2.2 Grouping Data

To understand most of a regression table, it will be important for us to group data in certain ways. We will use two of these groupings to estimate "population" variances. "Population" denotes all possible responses at the experimental levels. The data we collect in experiments is normally just a "sample" of that population. If we repeated the same experiment, we would probably collect a different "sample" from that same population. We are examining variance because there are some statistical tools at our disposal that lend themselves directly to comparisons of variances of samples from the same population. This will lead to judgements of significance. For simplicity, let's suppose we had only one factor at 2 levels, and eight runs that graphically looked like the following figure.

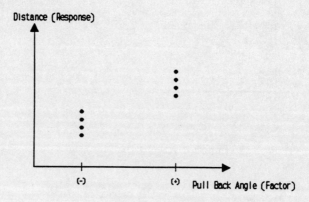

Figure 4.8 One Factor at Two Levels

Our prediction equation $\hat{y}$ provides a best fit for the data.

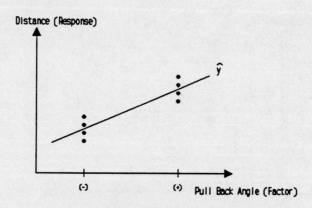

Figure 4.9 Fitting an Equation to Data

Overlaying the mean of the data points $\bar{\bar{y}}$,

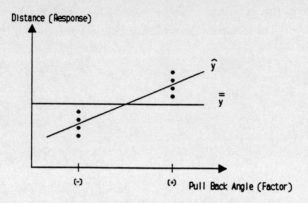

Figure 4.10 Overlaying $\bar{\bar{y}}$

we can begin to look at the variance in the data, and using that variance, estimate the variance of the all possible (population) responses in this factor range. There are three ways to estimate this variance.

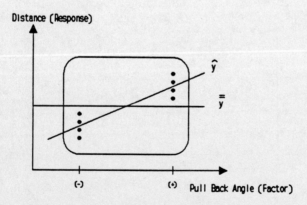

Figure 4.11a Estimate population variance from the variance of the entire data group about $\bar{\bar{y}}$.

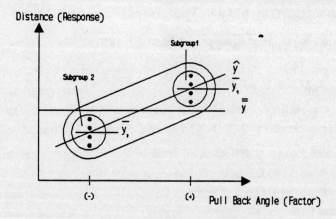

Figure 4.11b Estimate population variance by somehow pooling the variances of each of the subgroups about its own mean. (Variance **within** subgroups.)

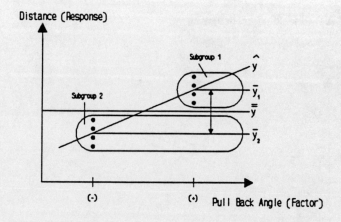

Figure 4.11c Estimate population variance by finding the variance of the mean of each subgroup of data about the grand mean $\bar{\bar{y}}$. (Variance **between** subgroups.)

We will use estimates 2 and 3 to judge significance of the model and the significance of individual factors (for a two-level design).

4.2.3 Standard Error of the Estimate:

($\sqrt{\text{variance within subgroups}}$ or Mean Square)

Statisticians tell us that we should expect our responses to fall normally about the prediction line with the same variance as the population. When we estimate the variance of the population with the variance within subgroups, we will call this the mean square within or mean square error (MSE). The population standard deviation will be estimated as $\sqrt{\text{MSE}}$, also known as the standard error of the difference (or estimate). This is found by summing the square of the difference between each data value and the predicted value for that factor level (**Sum of Squares Residual (or error) (SSE)**), and then dividing by a correction factor called degrees of freedom (df) which will be discussed later. In our example, this sum is 1298.333 and the df = 16. That means that the standard error for the estimate = $\sqrt{\text{MSE}} = \sqrt{\dfrac{1298.333}{16}} = 9.008$. Using the empirical percentages for a normal distribution, we expect 68.26% of any further statapult firings to fall inside of $\hat{y} \pm \sqrt{\text{MSE}}$ (one standard deviation).

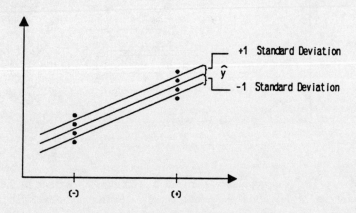

Figure 4.12 Interpretation of MSE

In our statapult example, we have three factors (A, B, and C) and one response (downrange distance, y). This is a four dimensional space that will be a little hard to visualize. Basically, $\sqrt{MSE}$ represents a distance above and below a planar surface that lies in this space. The three data points collected in a run represent a data subgroup. We have eight data subgroups. If this were a three dimensional space, we would picture these data subgroups floating above/below the appropriate settings for A and B (see Figure 4.13). (The $\sqrt{MSE}$ would represent a distance above and below the surface that we would expect 68.26% of all future values to fall within. The surface will be known as the "Response Surface.")

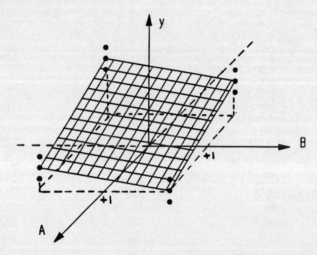

Figure 4.13 A Response Surface

For this "best" estimate of the population variance (MSE), we said that we would somehow pool the variance of subgroups. Mechanically, we said we would compute a sum of squares and divide by degrees of freedom. For our example, the computations are shown in Table 4.7.

Although it seems hidden in the mechanics, we have actually pooled the variances from each run. MSE is an overall measure of variation. Each subgroup is the set of data collected at a run setting. Since the predicted value for each data subgroup is the mean (for a two-level design), let's examine the SSE for one run:

$$SSE_{RUN1} = (85 - \bar{y}_{RUN1})^2 + (74 - \bar{y}_{RUN1})^2 + (58 - \bar{y}_{RUN1})^2$$

If we divide this by the number of data points for that run (n_{RUN}) minus one, you will see the familiar form of the variance for that run:

$$\frac{SSE_{RUN1}}{n_{RUN1}-1} = \frac{(85 - \bar{y}_{RUN1})^2 + (74 - \bar{y}_{RUN1})^2 + (58 - \bar{y}_{RUN1})^2}{3 - 1} = S^2_{RUN1}$$

Table 4.7 Computation of SSE

Run #	$(y_i - \hat{y})^2$ (Data value at run setting − Predicted value at that setting)²			
1	(85	−	72.33)² =	160.43
1	(74	−	72.33)² =	2.78
1	(58	−	72.33)² =	205.46
2	(128	−	127)² =	1.00
2	(132	−	127)² =	25.00
2	(121	−	127)² =	36.00
3	(150	−	138.5)² =	132.25
3	(145	−	138.5)² =	42.25
3	(120.5	−	138.5)² =	324.00
4	(215	−	209)² =	36.00
4	(208	−	209)² =	1.00
4	(204	−	209)² =	25.00
5	(32.5	−	35.67)² =	10.02
5	(36	−	35.67)² =	.112
5	(38.5	−	35.67)² =	8.03
6	(69	−	72.33)² =	11.10
6	(80	−	72.33)² =	58.8
6	(68	−	72.33)² =	18.77
7	(68	−	72.67)² =	21.79
7	(81.5	−	72.67)² =	78.00
7	(68.5	−	72.67)² =	17.37
8	(114	−	108.67)² =	28.45
8	(101.5	−	108.67)² =	51.35
8	(110.5	−	108.67)² =	3.36
			SSE =	1298.33

Proceeding in that fashion, we could find the SSE for a two-level design simply by finding the variance of the responses for each run, multiplying each by ($n_{RUNi} - 1$), and summing them:

$$SSE = (n_{RUN1} - 1)S^2_{RUN1} + (n_{RUN2} - 1)S^2_{RUN2} + (n_{RUN3} - 1)S^2_{RUN3}$$
$$+ (n_{RUN4} - 1)S^2_{RUN4} + (n_{RUN5} - 1)S^2_{RUN5} + (n_{RUN6} - 1)S^2_{RUN6}$$
$$+ (n_{RUN7} - 1)S^2_{RUN7} + (n_{RUN8} - 1)S^2_{RUN8}$$

$$= \sum_{i=1}^{\#RUNS} (n_{RUNi} - 1)S^2_{RUNi}$$

In our example,

$$SSE = (3 - 1)(184.33) + (3 - 1)(31.00) + (3 - 1)(249.26) + (3 - 1)(31.00)$$
$$+ (3 - 1)(9.08) + (3 - 1)(44.33) + (3 - 1)(58.58) + (3 - 1)(41.59)$$
$$= 1298.33$$

4.2.4 Degrees of Freedom

Since the **degrees of freedom (df)** have entered our computations, perhaps now is the time to discuss them. The best explanation we have encountered is in the following computation of a mean. Suppose you know that the mean of five numbers is 25. How many free choices do you have in selecting the numbers that will make this happen? We propose that the first four numbers are up to you and that the fifth is predetermined by the mean. Therefore, you have four degrees of freedom. In a similar manner, we had three data points in each run to compute a S^2_{RUN}. This gives us two degrees of freedom for each of our runs and a total of 16 (2 df × 8 runs) for the SSE. Therefore, MSE for two-level factors is:

$$MSE = \frac{\sum_{i=1}^{\#RUNS} (n_{RUNi} - 1)S^2_{RUNi}}{\sum_{i=1}^{\#RUNS} (n_{RUNi} - 1)}$$

4.2.5 Mean Square Between/Regression

The remainder of our analysis is a comparison of the population variance estimate between subgroups (Mean Square Between (MSB) considering all factors, then each separately) with the Mean Square Error. Take another look at the variance between subgroups graph on page 4-19 and Figure 4.14. It seems reasonable to state that the only time the "between estimate" will be close to that of the standard error of the estimate will be when there is very little "shift" in the subgroups about $\bar{\bar{y}}$, e.g.

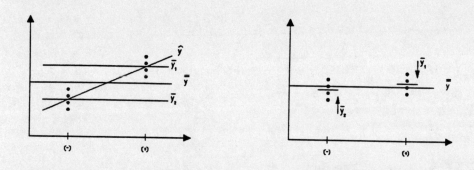

Figure 4.14 MSB (Shifts in Subgroup Means)

If there was very little shift in response between levels of all factors, then the prediction equation would not "predict" much better than the mean. For each factor's influence considered separately, this would indicate that the factor would have little influence on the response. So, in order for the model to be a good predictor or for a factor to influence the response, the variation between subgroups must be somewhat bigger than the MSE. (Somewhat bigger will be defined by the t, z, or F test as a certain measure of confidence that the model has detected a significant shift in the mean, or that a factor should be included in the model.)

The computations of the MSB are as straightforward as those of the MSE. When we look at this "between" estimate for the overall model we will refer to it as the Mean Square Regression (MSR). For individual factors we will use MSB. Mechanically, we are looking at a variance based on the difference between the predicted value for a specific run setting ($\hat{y}$) and the overall mean ($\bar{\bar{y}}$). This will lead us to the MSR. For each factor's MSB, we will examine only the predicted value at each factor level and compare that with $\bar{\bar{y}}$.

First, let's attack the MSR. For our example, we will compute a sum of squares and then adjust it with the degrees of freedom we can attribute to the regression model

(Table 4.8).

Table 4.8 Sum of Squares Regression

Run #	$(\hat{y} - \bar{\bar{y}})^2$ (Predicted Value at run/factor settings − Grand Mean)²				
1	(72.33	−	104.521)²	=	1036.26
1	(72.33	−	")²	=	1036.26
1	(72.33	−	")²	=	1036.26
2	(127	−	")²	=	505.31
2	(127	−	")²	=	505.31
2	(127	−	")²	=	505.31
3	(138.5	−	")²	=	1154.57
3	(138.5	−	")²	=	1154.57
3	(138.5	−	")²	=	1154.57
4	(209	−	")²	=	10915.86
4	(209	−	")²	=	10915.86
4	(209	−	")²	=	10915.86
5	(35.67	−	")²	=	4740.46
5	(35.67	−	")²	=	4740.46
5	(35.67	−	")²	=	4740.46
6	(72.33	−	")²	=	1036.26
6	(72.33	−	")²	=	1036.26
6	(72.33	−	")²	=	1036.26
7	(72.67	−	")²	=	1014.49
7	(72.67	−	")²	=	1014.49
7	(72.67	−	")²	=	1014.49
8	(108.67	−	")²	=	17.21
8	(108.67	−	")²	=	17.21
8	(108.67	−	")²	=	17.21
			SSR	=	61261.906

Once again, the computations have partially hidden what we are doing. For each run you have probably noticed that the predicted value was reported three times − one for each data point taken during that run. Therefore, the sum of squares is nothing more than the sum of the number of data points collected during each run times the squared difference between the predicted value for a run and the overall mean.

To use this sum of squares to estimate a population variance, we still need to adjust it with the degrees of freedom. Since our data points for MSR are the predicted values for each run setting (for a two-level design, this will be the average of the data values for a run), we have 8 data points. The degrees of freedom will be 7 for the regression model. **Notice that this is also the number of variable parameters in our model (A, B, C, AB, AC, BC, ABC).** Thus,

$$MSR = \frac{SSR}{df} = \frac{61261.906}{7} = 8751.701$$

Although the MYSTAT regression model doesn't examine MSB for each factor, let's tie that in. Recall the scatterplot for the pullback angle (A) vs distance shown in Figure 4.15. If we include the overall mean and the average at the high and low levels, we can begin to understand the MSB for factor A. Just like we did for MSR, we will calculate a sum of squares:

$$SS_{FACTOR\,A} = \binom{\text{\# of data points}}{\text{at low level}}(\bar{y}_- - \bar{\bar{y}})^2 + \binom{\text{\# of data points}}{\text{at high level}}(\bar{y}_+ - \bar{\bar{y}})^2$$

However, you know that the difference between $\bar{y}_-$ and $\bar{y}_+$ is the Δ for factor A (ΔA). This also tells you that if we split the difference with $\bar{\bar{y}}$, $\bar{y}_- - \bar{\bar{y}} = \bar{y}_+ - \bar{\bar{y}} = \Delta/2$. Our sum of squares now looks somewhat simpler:

$$SS_{FACTOR\,A} = \binom{\text{\# of data points}}{\text{at low level}}\left[\frac{\Delta}{2}\right]^2 + \binom{\text{\# of data points}}{\text{at high point}}\left[\frac{\Delta}{2}\right]^2$$

$$= (\Delta/2)^2(\text{total \# of data points})$$
$$= (\Delta/2)^2 N$$
$$= \Delta^2(N/4)$$

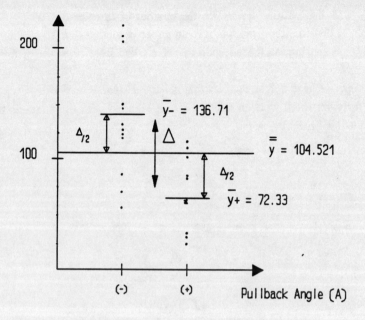

Figure 4.15 Scatterplot for Pull Back Angle

Now, we must adjust this sum of squares to estimate the population variance. Once again we will divide by the degrees of freedom. For MSB, our two means (high and low) are the only two data points. Using our previous argument, the degrees of freedom will be one. Therefore, the

$$\text{MSB}_{\text{FACTOR A}} = \frac{\frac{\Delta^2 N}{4}}{1} = \frac{\Delta^2 N}{4} = \frac{(-64.37)^2 (24)}{4}$$

$$= 24,864.07$$

The simpler formula for $\text{MSB}_{\text{FACTOR}} = \Delta^2 N/4$ will always be valid for each factor in a **two-level orthogonal design** with an equal number of data points for each row.

4.2.6 Distributions

Now that we have all these estimates for variance, what do we do with them? To answer this question, we need to review three types of distributions:

1. Z distribution (used when the number of samples ≥ 30)
2. t distribution (used when the number of samples < 30)
3. F distribution (based upon samples taken from normal distributions)

In our seminar we have a little exercise that helps our students visualize what is going on. We have forty-six small wooden balls in a sack labelled with the numbers 0-11. They are distributed as in Figure 4.16.

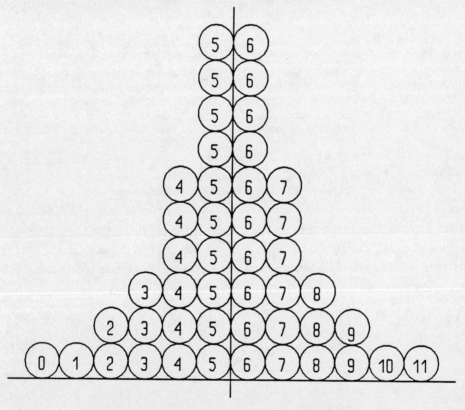

Figure 4.16 Approximate Normal Distribution

If we overlay the "bell-shaped" function, it becomes quite clear that our distribution is approximately "normal." The Z and t distributions fall into this category. The t distribution is a little flatter, based on the number of samples in the distribution. Most commonly, these distributions are standardized with a mean of zero, a standard deviation of one, and such that the area under the curve is one. That tells us that the horizontal axis will be the number of standard deviation units we are on either side of zero and the vertical axis a measure of "frequencies." Usually we will be interested in the probability of being so many standard deviation units above or below zero. This is computed as area under those curves. For example, if we wanted to know the probability of being 1.6 standard deviation units to the right of zero on the Z distribution we would compute the area under the curve from $-\infty$ to 1.6. The good news is that these values are given to us by computers or tables.

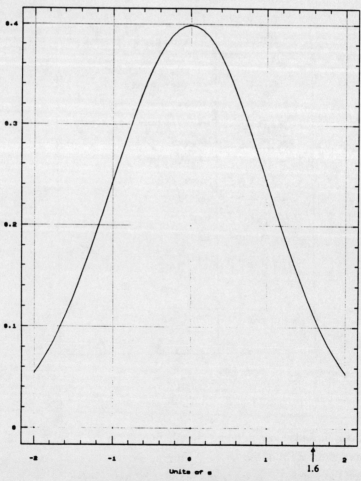

Figure 4.17 Z Distribution

4.2.7 F Statistic

We have each student take five balls from the sack, record the numbers, and put them back in the sack. Then we ask them to compute the variance of their five numbers and divide it by the variance computed by two or three other students. Afterward we have each student give us their ratios. A typical count of the ratios looks like:

Table 4.9 Empirical F Distribution

Ratio	Number in Ranges
0-0.5	**************
>0.5-1.0	***************
>1.0-1.5	************
>1.5-2.0	******
>2.0-2.5	******
>2.5-3.0	****
>3.0-3.5	**
>3.5-4.0	***
>4.0-4.5	**
>4.5-5.0	***
>5.0-5.5	**
>5.5-6.0	*
>6.0	***

Empirically, our students have built an F distribution — the graph being visible when we overlay a curve to the right of the "stars." Turning the picture more conventionally, the horizontal axis represents the F ratio and the vertical axis a measure of frequency.

Analyzing Your Results 4-31

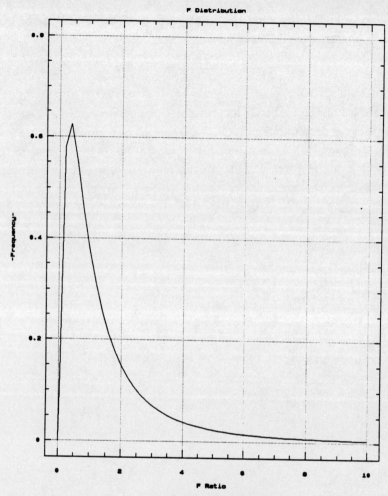

Figure 4.18 F Distribution

Although the "F distribution" will change with the degrees of freedom of the numerator and denominator of the F ratio, this gives us a basis for understanding what is going on.

What have we learned? If we compared the variance of samples taken from an approximate normal distribution (obviously the samples came from populations with the same mean), we see that most of the ratio values cluster around one. Does that seem intuitive? And very few are greater than six. The chance of finding a specific F value is found in a manner similar to the chance of being any number of standard deviation units from zero in the Z distribution. The area under the F curve will also be one.

Let's look at the overall F ratio for our model:

$$F = \frac{MSR}{MSE} = \frac{8751.70}{81.146} = 107.852$$

(We will also place our best estimate for variance in the denominator.)

We saw in our demonstration that $F > 6.0$ seemed rare. In fact, we use $F = 6.0$ as our Rule of Thumb [1] for a cut-off. If $F > 6.0$, we will say that there is a significant shift in the response of different run settings. That means that we don't believe the change in response at different settings happened by chance; we believe it

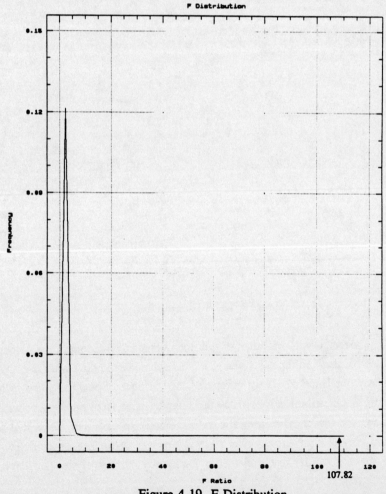

Figure 4.19 F Distribution

happened because there is a difference in responses between factor levels.

To try to get a handle on this "chance," let's look again at a picture of an F distribution (Figure 4.19). Based on our previous discussion and demonstration, we believe that 107.852 is a large F value — located considerably far out on the "tail" of the distribution. Since we do not believe there is a high probability that would happen by chance, we are saying we think that happened because the factors shifted the response. Essentially this says that we do not believe the two estimates of variance came from the same populations. We think that the response actually behaves differently (has different averages) for different factor levels.

4.2.8 Type I Error (α)

Since there is a little bit of the "tail" to the right of 107.852, there appears to be some chance we might be wrong that the shift in response is due to a change in factor levels. In other words, there is a small probability $F = 107.852$ could have happened by chance. This chance or risk is known as a type I error (α error).

The best example we have seen that explains an α error is the decision reached by a jury regarding a defendant [2]. The following table summarizes the possibilities:

Table 4.10 Type I Error

		Reality	
		Not Guilty	Guilty
Verdict	Not Guilty		Type II (β) Error
	Guilty	Type I (α) Error	

If the jury appropriately finds an innocent defendant innocent or a guilty defendant guilty, we have no disagreement. In our judicial system we guard against finding an innocent person guilty. We consider that the worst possible error. This type error is called a type I error (α). The other situation is a type II (β) error. We will only concentrate on the α error. We will choose α = .05 (telling us we will be 95% confident that our decision is correct). This selection is arbitrary.

Relating this to our problem, let's suppose we want to be 95% (α = .05) confident that our model detects a linear shift in response due to changes in factor levels. Then we will want the area under the F curve to be greater than or equal to .95 for our F = 107.82.

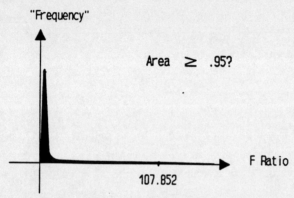

Figure 4.20 Area Under "F" Probability Distribution Function

4.2.9 P Value

The P value is the area remaining in the tail. In the analysis of variance table we see that P = 0.000. That tells us that the area to the right of 107.82 is pretty close to zero. In other words, we can be at least 99.9% confident that our model detects a shift in data.

We are getting a similar piece of information in the regression table with this P value based on the t distribution. (Note that in 2-level designs the t statistic is $\sqrt{F}$ statistic. In a two-level design, we can calculate the F statistic for each factor using $F = \dfrac{MSB_{FACTOR}}{MSE} = \dfrac{\frac{N\Delta^2}{4}}{MSE}$ and then compute $t = \sqrt{F}$. Recall that for factor A, MSB = 24864.07. Calculating F, $F = \dfrac{24864.07}{81.146} = 306.4115 \Rightarrow t = 17.505$. We will discuss the sign difference momentarily. From our rule of thumb, this factor appears significant. The P value supports that conclusion.)

If P > .10 we recommend you not include that factor in your model.

4.2.10 T Statistic/Standard Error of the Coefficient

Since all of your designs will not be two level, we should spend some time explaining the regression table in more detail. Let's not forget that our objective is to determine which factors and interactions belong in our model and which do not. If a factor does not significantly affect the response, then we should expect that factor not to appear in our model. In other words, we should expect the coefficient of that factor to be zero. So, our analysis will dwell on whether a particular coefficient is large enough to include it in the model. This is essentially the same question we are faced with when we draw a Pareto Chart of our half effects in a two-level design — particularly since the half effects are the coefficients.

If we were to run the entire experiment over again, we would get a different set of 24 data points. That would ensure a slight difference in most of our half effects. If we did this many times, we would eventually see a distribution of each half effect. If a factor had no effect on a response, we should expect these half effects to cluster around zero. Therefore, we will make a probability judgement about the coefficient in the same way we did about the size of the F ratio. Since the coefficient may be positive or negative and we expect the half effects to be distributed normally, we will use the t distribution. Specifically, let's concentrate on the coefficient for factor A (pullback). If the value is labeled on a t distribution (see Figure 4.21 on the following page), the question really becomes, "How far out on the tail is this value compared to all others?" To answer this question, we will "standardize" the coefficient values and use standard tables. Standardizing simply means to write the coefficients in terms of standard deviation units. Your concern here is simply how does the coefficient vary?

This is not a straightforward answer since we need to determine how each coefficient is calculated. The matrix algebra computations are included as Appendix B. From Appendix B, the coefficients determined from a two-level orthogonal design vary identically:

$$\sigma^2(b_0) = \sigma^2(b_1) = \ldots = \sigma^2(b_a) = \frac{\sigma^2(y)}{n} \approx \frac{MSE}{n}$$

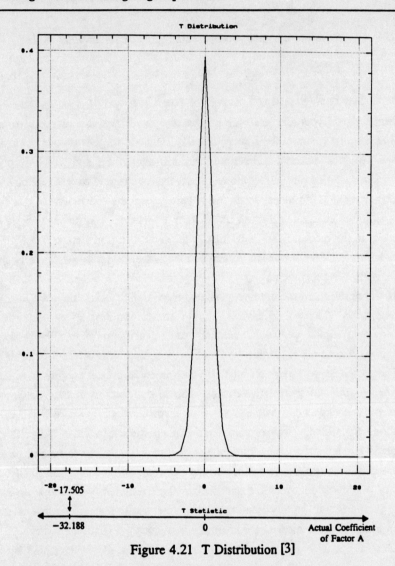

Figure 4.21 T Distribution [3]

The STD ERROR (of the coefficients) is then determined as:

$$\sigma(b_0) = \sigma(b_1) = \ldots = \sigma(b_n) \approx \sqrt{\frac{MSE}{n}} = \frac{\text{STANDARD ERROR OF THE ESTIMATE}}{\sqrt{n}}$$

Now we can state precisely the number of standard deviation units each coefficient is from zero. For the coefficient of factor A, -32.188 is $\frac{-32.188}{1.839}$ or -17.5049 standard deviation units to the left of zero on the t distribution. The P value tells us that 99.9% of the values fall within ± 17.5049. We believe we are far enough out on the tail (or 99.9% confident) that the coefficient is significantly different from zero.

4.2.11 Standard Coefficient

Because each factor may represent quite different types and units of measurement (e.g. PSI vs. °F), it is sometimes hard to judge sensitivity. This standardization makes the coefficients unitless and reflects the change in mean response per unit change in the factor (both in their respective standard deviation units). [4]

4.2.12 Multiple R: .990

The correlation coefficient (R) for multiple regression. Square root of the coefficient of determination (R^2). Closer to ± 1 is better but **caution**, because R close to zero may not be bad. (See Squared Multiple R.)

4.2.13 Squared Multiple R: .979

The coefficient of determination (R^2) for multiple regression. In this case, $R^2 = .979$ tells us that 97.9% of the variance in the data (about the mean) was due to the fact that the response varied for different levels of the factors. It also tells us that 2.1% of the variance was simply due to the variance of responses at each level. Let's use our example and compute these:

y_i = data points (we have 24)
$\hat{y}_i$ = predicted value for y at a particular factor setting
$\bar{\bar{y}}$ = grand mean

We know that variance is calculated by summing the square of the difference between each data point and the grand mean $\bar{\bar{y}}$ ⇒ Total Variance = $\dfrac{\sum_{i=1}^{n}(y_i - \bar{\bar{y}})^2}{n-1}$.

In order to relate this to the prediction equation, we have the following equations.

$$y_i - \bar{\bar{y}} = y_i - \hat{y} + \hat{y} - \bar{\bar{y}}$$

$$y_i - \bar{\bar{y}} = (y_i - \hat{y}) + (\hat{y} - \bar{\bar{y}})$$

These are depicted graphically in Figure 4.22. The difference between the predicted response at a setting and a data value at that setting ($y_i - \hat{y}$) is the residual.

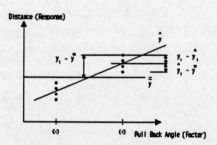

Figure 4.22 Residuals

If we square both sides and sum for all data responses, we get (after a lot of math):

$$\Sigma(y_i - \bar{\bar{y}})^2 = \Sigma(y_i - \hat{y})^2 + \Sigma(\hat{y} - \bar{\bar{y}})^2$$

Recall earlier that $\Sigma(y_i - \hat{y})^2$ was the Sum of Squares due to Error (SSE) and $\Sigma(\hat{y} - \bar{\bar{y}})^2$ was the Sum of Squares due to Regression (SSR). That makes $\Sigma(y_i - \bar{\bar{y}})^2$ the Sum of Squares Total (SST). If we divide both sides by n−1, the left hand side is the total variance with the other two pieces of the equation being the variance due to error and the variance due to regression.

$$\dfrac{\sum_{i=1}^{n}(y_i - \bar{\bar{y}})^2}{n-1} = \dfrac{\sum_{i=1}^{n}(y_i - \hat{y})^2}{n-1} + \dfrac{\sum_{i=1}^{n}(\hat{y} - \bar{\bar{y}})^2}{n-1}$$

Dividing through by total variance,

$$1 = \frac{\sum (y_i - \hat{y})^2}{\sum (y_i - \bar{\bar{y}})^2} + \frac{\sum (\hat{y} - \bar{\bar{y}})^2}{\sum (y_i - \bar{\bar{y}})^2}$$

SSR/SST = R^2 (the coefficient of determination).

$$\rightarrow 1 = \frac{\sum (y_i - \hat{y})^2}{\sum (y_i - \bar{\bar{y}})^2} + R^2$$

If we consider the following two cases, we will understand why 0 and 1 are our bounds.

CASE 1: If we are able to fit our line through all data points, $y_i = \hat{y}$ at each x_i and
$1 = 0 + R^2$
$1 = R^2$
which means that 100% of the variance about the mean is simply due to a shift in responses at different factor levels - that's good!

CASE 2: If the best predictor for our data is the mean ($\hat{y} = \bar{\bar{y}}$), then
$1 = 1 + R^2$
$0 = R^2$
which tells us that we can detect no variance in the data due to shifts in responses about the mean for different factor levels. In other words 100% of the variance is due to error in our model not fitting the data.

In our example, we have already calculated the SSE and the SSR. The sum of squares total could be computed identically, but is also the sum of SSE and SSR.

Variation due to Error and Regression can be calculated as follows:

$$\frac{SSE}{SST} = \frac{1298.333}{62560.239} = .021$$

This tells us that 97.9% of the variation is due to the shift in responses for different levels

$$\frac{SSR}{SST} = (R^2) = \frac{61261.906}{62560.239} = .979$$

of the factors. Some books refer to this as a percent of the variation accounted for by y's linear relationship with x.

Many people treat R^2 as a "goodness of fit" number, with a higher R^2 indicating a better fit. That seems to imply that a predictive "surface" fits the given data better if R^2 is closer to 1.0. That may not be true. We prefer to use R^2 as a measure of how well our model can account for shifts in the average response for different factor levels. If R^2 is high, we are more confident that we have found a relationship between the factors and the response and that more of the variation is explained. That is generally "good." If R^2 is low, we have not found a relationship between the selected factors (in the selected experimental ranges) and the response. Little of the variation is explained. That is generally "bad." That will tell us to hunt elsewhere — perhaps with different levels for our factors.

4.2.14 Adjusted Squared Multiple R: .970

This R^2 compensates for models with large numbers of terms compared to the number of observations. The formula is [4]:

$$\text{Adj } R^2 = 1 - \left[\frac{n-1}{n-p}\right](1 - R^2)$$

where n = total number of observations
p = total number of terms in the model (including the constant)

In our example,

$$\text{Adj } R^2 = 1 - \frac{(24-1)}{(24-8)}(1 - .979)$$
$$= 1 - \frac{23}{16}(1 - .979)$$
$$= .970$$

Basically, a large number of terms for a relatively small n may lead to an overfit condition. This will mask the true variation due to error. Adj R^2 inflates the error estimate using a ratio of degrees of freedom and recomputes R^2.

4.2.15 Tolerance

Tolerance is calculated for each factor. A regression is run for a particular factor in terms of all remaining factors. Then an R^2 is calculated for that regression. Tolerance is equal to $1 - R^2$ (for that regression).

For example, the regression equation for factor A would be A = constant + (coefficient)B + (coefficient)AB + all other factors. If there is no shift in A due to different levels of the other factors and interactions, $R^2 = 0$ and the tolerance for that factor = 1. A tolerance of 1.00 suggests orthogonality.

4.3 Summary

The following chart summarizes the items on a regression output table:

Regression Name	Meaning
Dep Var:	The dependent variable — the name of your response.
N:	The number of data points.
Multiple R:	The correlation coefficient (R) for multiple regression. Closer to ± 1 is better.
Squared Multiple R:	The correlation coefficient (R^2). A measure of the relationship between the factors and the response. Closer to 1 indicates that there is a relationship and more variation is explained.
Adjusted Squared Multiple R:	R^2 adjusted for the number of terms in the model when compared with the number of data points. Adjustment for an overfit condition.
Standard Error of Estimate:	Expected deviation of data about the predictive "surface." $\sqrt{MSE}$.
Variable:	The terms (factors) in the predictive model.
Coefficient:	The multiplier for each term in the model.

Std Error:	Expected deviation of each coefficient if more data were collected.
Std Coeff:	Adjusted coefficient — makes it dimensionless.
Tolerance:	A measure of correlation between the individual term in the model and all others. Orthogonal designs will provide a tolerance of 1 for each term.
T:	The "t" statistic. The number of normalized standard deviation units the coefficient is from zero. The further from zero the coefficient, the more confidence we have in its effect on the response.
P(2 tail):	The probability the term should **NOT** be in the model. If P > .10, consider excluding it from the model.
Sum-of-Squares Regression:	The sum of the squared difference between the predicted value for each run and the overall average response $$\left(\sum_{i=1}^{N} (\hat{y}_i - \bar{\bar{y}})^2\right) \text{ (SSR)}.$$
Sum-of-Squares Residual (Error):	The sum of the squared difference between each data point and the predicted value for that data point $$\left(\sum_{i=1}^{N} (y_i - \hat{y}_i)^2\right) \text{ (SSE)}.$$
DF:	Degrees of freedom. The number of free choices we have in determining a value.
DF Regression:	The number of non-constant terms in the model.
DF Residual:	N + 1 − (DF Regression)
Mean-Square Regression	"Between" estimate of variance for the overall model (MSR). MSR = SSR/DF$_{REGRESSION}$.
Mean-Square Residual (Error):	"Within" estimate of variance. Best estimate of population variance (MSE). MSE = SSE/DF$_{REGRESSION}$.
F-Ratio:	"F" statistic (F = MSB/MSE). A higher number indicates the model detects a relationship between the factors and the response.
P:	Probability that this is **NOT** a good model. "Good" indicates that a relationship between factors and response has been found.

Based upon our analysis, only the effects of BC and ABC are considered unimportant. Therefore, our prediction equation for the mean response is:

$$\hat{y} = 104.521 - 32.188A + 27.688B - 9.354AB + 24.729C - 6.563AC$$
(We normally want to include an equation for $\hat{s}$ or ln $\hat{s}$, but you will recall that no factor significantly affected the variance.)

Before we assume our prediction equation is correct, let's predict the settings required to put our process on a target and then do a few experimental confirmation runs to gain confidence that our prediction equation is useful.

Chapter 4 Bibliography

1. Schmidt, S. R. and Launsby, R. G., *Understanding Industrial Designed Experiments (3rd edition)*. Colorado Springs, CO: Air Academy Press, 1991.

2. Kiemele, M. J. and Schmidt, S. R., *Basic Statistics: Tools for Continuous Improvement (2nd edition)*. Colorado Springs, CO: Air Academy Press, 1991.

3. *Statgraphics (Version 5)*, STSC, Inc., Rockville, MD.

4. Neter, John; Wasserman, William; and Kutner, Michael H., *Applied Linear Statistical Models (2nd edition)*. Homewood, IL: Richard D. Irwin, Inc., 1985.

CHAPTER 5
PUTTING YOUR PROCESS ON TARGET
(with Minimum Variation)

Now that we have prediction equations, what do we do next? This is the part we have been waiting for — the reason we did all this work! We can put our process on target, hopefully with minimum variation.

Be cautious as you go because:

YOUR PREDICTION EQUATION IS ONLY VALID IN YOUR EXPERIMENTAL REGION. IF YOUR TARGET VALUE (whether maximum, minimum, or somewhere in between) DOES NOT LIVE IN THIS REGION, YOUR PREDICTION WILL VERY LIKELY BE INVALID!

Even if our target value falls outside our experimental region, that does not mean our experiment is a failure. It may actually point us in the right direction to continue searching.

There are a few ways we can proceed:
1. Use the settings, indicated by the plot of averages, that maximize or minimize the response and/or variation.
2. Set the prediction equation equal to your target value and "play" with the settings for each factor until you get close. Most software packages have a utility that allows you to experiment with the settings to achieve a desired prediction.
3. Find the partial derivatives of our prediction equations with respect to each factor in the equation. Set those equal to zero and solve. The answers will be those settings that maximize or minimize the prediction, or indicate a

"saddle point." The second derivative test will confirm which it is.
4. Use a response surface method (RSM). This is a class of methods that uses the idea that the prediction equation defines a "surface." For one or two factors, this will be clear. For three or more, it may not be so obvious.
   a. **Contour Plot:** Here we will use our surface to define the lines of constant response and graphically project those lines onto a two dimensional graph. Then we will be able to see those factor settings that give us a specific response.
   b. **Method of Steepest Ascent/Descent:** This is a numerical method that uses the gradient of the prediction equation to point us toward a maximum or minimum. (This method will not be discussed further, but may be reviewed in the text, *Response Surface Methodology*, by Raymond H. Myers.)

Let's use our catapult data to put our process on target. Our response equation was:

$$\hat{y} = 104.521 - 32.188A + 27.688B - 9.354AB + 24.729C - 6.563AC$$

5.1 Apply Settings from Plots of Averages

Recall from our plot of averages, we will reach our maximum distance when A is set $(-)$, B is set $(+)$, and C is set $(+)$:

$$\hat{y} = 104.521 - 32.187(-1) + 27.688(+1) - 9.354(-1)(+1) + 24.729(+1)$$
$$- 6.563(-1)(+1)$$
$$= 205.042 \text{ inches}$$

(Note that all the settings, A_-, B_+, and C_+ are within the experimental region.)

We can find the minimum distance in a similar manner. If A is set $(+)$, B is set $(-)$, and C is set $(-)$:

$$\hat{y} = 104.521 - 32.188(+1) + 27.688(-1) - 9.354(+1)(-1) + 24.729(-1)$$
$$- 6.563(+1)(-1)$$
$$= 35.833$$

(Once again, all factor settings are within the experimental region.)

5.2 Set Prediction Equation Equal to Target and Find Settings

It should be obvious that we can choose a target somewhere between 35.833 and 205.042 inches and find settings that get us close to that target. For example, let's find the settings that will target the in-flight distance at 100 inches. Without doing anything it might have occurred to you that setting $A = B = C = 0$ will get us pretty close. However, we must not forget that our settings for B and C are limited by placement of the holes on the catapult.

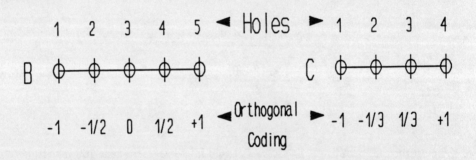

Figure 5.1 Orthogonal vs. Real Settings for Hook and Peg

If we set $B = 0$ and $C = 1/3$, we can "play" with A until we get close to 100 inches.

$$\hat{y} = 104.521 - 32.188A + 27.688(0) - 9.354(A)(0) + 24.729(1/3)$$
$$- 6.563(1/3)(A)$$
$$= 112.764 - 34.376A$$

When $\hat{y} = 100 = 112.764 - 34.376A$
$$\Rightarrow 34.376A = 12.764$$
$$A = .371$$

5-4 Straight Talk on Designing Experiments

And to convert back to a real setting

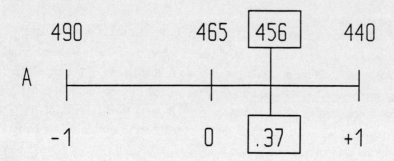

$$A_R = \frac{A_{CODED} \times d_{REAL}}{Z} + \bar{A}_R$$

$$= \frac{(.371)(-50)}{2} + 465$$

$$\approx 456$$

So, if we set the pull back angle at 456, we should hit somewhere near 100 inches. We tried it in our office and hit 102 and 99 inches! We will cover more about confirmation in Chapter 6.

5.3 Use Partial Derivatives to Locate Optima

For the maximum and minimum found in Section 5.1, we should take note that these are only the optimums within our region. Since our prediction is linear, do you suspect a real minimum or maximum? Using partial derivatives:

$$\frac{\partial \hat{y}}{\partial A} = -32.187 - 9.354B - 6.563C$$

$$\frac{\partial \hat{y}}{\partial B} = 27.688 - 9.354A$$

$$\frac{\partial \hat{y}}{\partial C} = 24.729 - 6.563A$$

After setting each equal to zero,

$$-32.187 = 9.354B + 6.563C$$
$$2.96 = \frac{27.688}{9.354} = A$$
$$1.94 = \frac{24.729}{6.563} = A$$

We see that there are an infinite number of settings for B and C that lead to $\partial \hat{y}/\partial A = 0$, and two settings for A that allow $\partial \hat{y}/\partial B$ and $\partial \hat{y}/\partial C = 0$. However, all of these settings are outside our region — and none of them lead to a real minimum or maximum. With a planar surface, there may be places where the plane is level with respect to a factor and yet not be an optimum. (Note that this flat part of the surface provides "robust" settings for those factors. This will be discussed in greater detail in Chapter 7.)

5.4 Contour Plots

Perhaps one of the most common response surface methods, the contour plot, will give us some insight here. Since $\hat{y}$ contains three variables, a two dimensional graph is difficult to draw. Normally, we will **arbitrarily** set the least significant factor (in this case C) at a least cost or nominal setting, e.g., $C_{ORTHOGONAL} = 1/3 \Rightarrow C_{REAL} =$ Hole #3). To give you an idea of how the plots change for different values of C, we have included both **response surfaces** and contour plots for C = 1/3, −1/3, and 1. Contour plots are simply projections of lines of constant response from the response surface onto the two dimensional "factor" plane. The two plots give us quick, valuable information.

5-6 Straight Talk on Designing Experiments

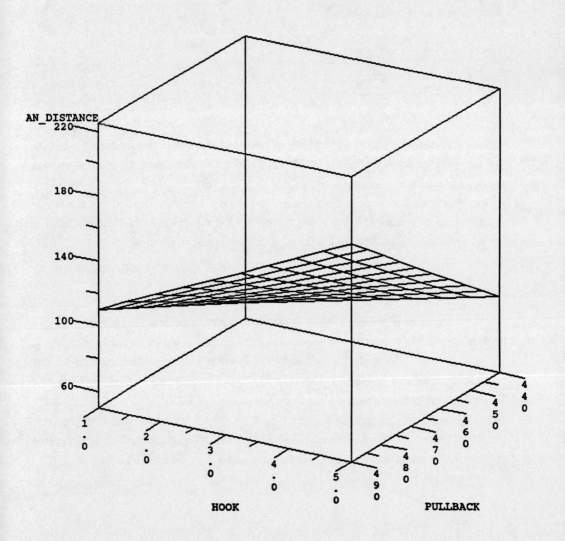

Figure 5.2 Response Surface for Statapult (C = 1/3)

The response surfaces for the C = −1/3 and C = +1 settings are:

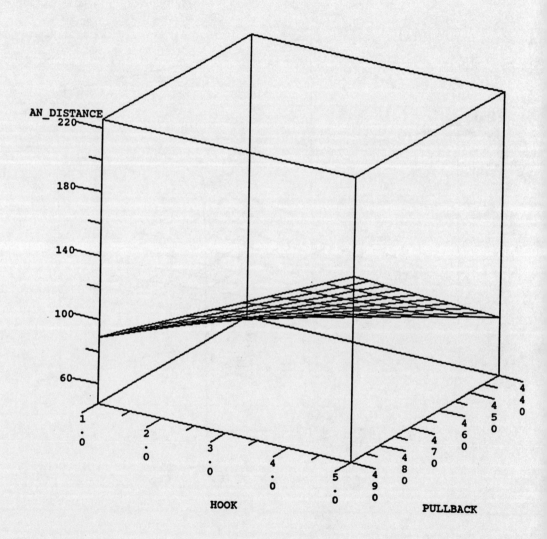

Figure 5.3 Response Surface for Statapult (C = −1/3)

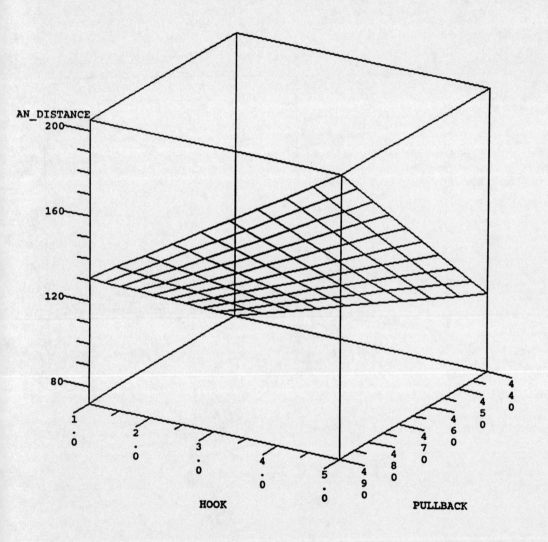

Figure 5.4 Response Surface for Statapult (C = 1)

By projecting the lines of constant distance down onto the horizontal (AB) plane, we create a contour plot. For C = 1/3, it looks like:

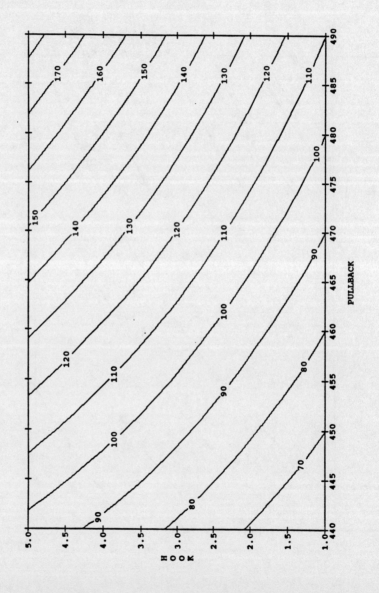

Figure 5.5 Contour Plot for Statapult (C = 1/3)

For $C = -1/3$ and $C = 1$, the contour plots are:

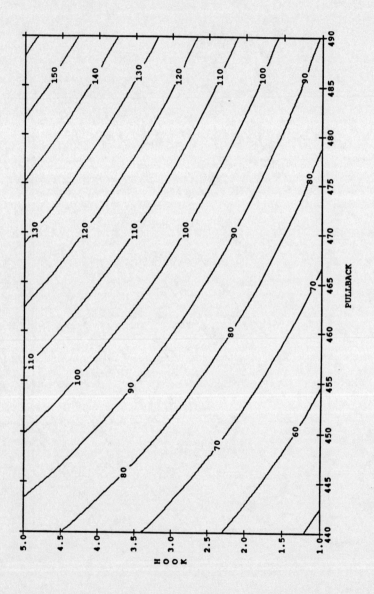

Figure 5.6 Contour Plot for Statapult ($C = -1/3$)

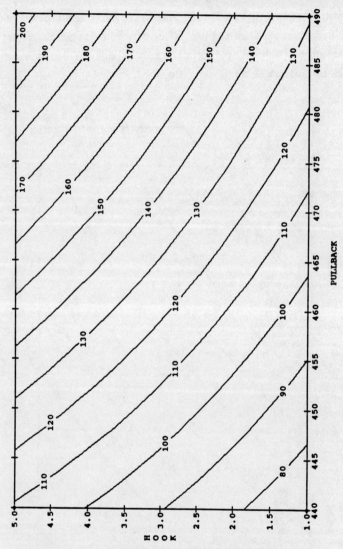

Figure 5.7 Contour Plot for Statapult (C = 1)

Using any of these plots, we can find a number of settings for the factors that allow us to hit a target of 100 inches. The horizontal and vertical axes represent two of the factors. The interior lines are lines of constant response. If our $\hat{s}$ equation had indicated any significant factors, then we would be able to use those to set the quantitative factors, B and C, at minimum variation settings. (Our current $\hat{s}$ "suggests" setting C at +1 will minimize variation. Without appropriate significance levels this is meaningless, but it will allow us to illustrate this method.) So, for C = +1,

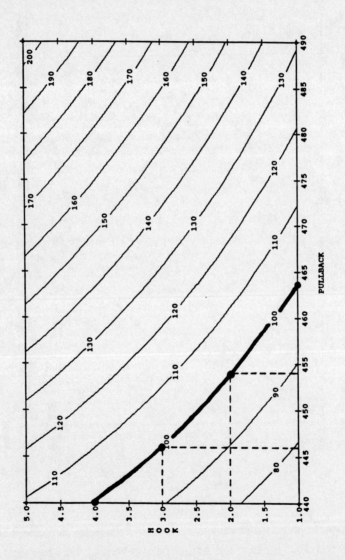

Figure 5.8 Using the Contour Plot to Target Our Statapult for 100 Inches

The following settings allow us to target 100 inches (Figure 5.8).

Real Values

A	B	C	ŷ
440	4	4	100
446	3	4	100
454	2	4	100
463.5	1	4	100

Essentially, this leaves the choice of settings for A and B up to you. Commonly you would try to choose based upon low cost or low variability.

Because of the software becoming available and the ease of interpretation, contour plots are quickly becoming the method of choice for many engineers. But there may be times when this will not work. If all of our catapult factors had been continuous or if we had entertained a fourth significant factor, the interpretation of a series of contour plots may not have been so straightforward. This is where a numerical method would be helpful, particularly if the location of an optimum is unknown. (Assuming, of course, that an optimum is desired). Even in our example, it appears that the optimum is not in our experimental region — but the contour plot does indicate where we might experiment next. In our case, we can do no more without drilling new holes or wrapping the rubber band tighter. We used all of our catapult's extreme settings.

Now, we CAN put our process on target and hopefully minimize the variation, too! If we were trying to improve a measure of quality, now is the time to check. With the process on target and variation reduced, are you better off now than you were before you started? Let's confirm!

Chapter 5 Bibliography

1. *RS Discover*, BBN Software Products Corporation, Cambridge, MA.

CHAPTER 6
CONFIRMATION

> **XI. DO CONFIRMATORY TESTS.**

Before we begin to use our new-found prediction equation to assess results or make decisions, we need to be sure our equations actually predict the results our process gives us for the same settings. For example, if we set factors A, B, and C to one set of levels indicated at the end of Chapter 5 that will hit a target of 100 inches:

Factor	Real Setting	Orthogonal Setting
A	454	+.44
B	Hole #2	−0.5
C	Hole #4	+1

the predicted value of our orthogonal equation is:

$$\hat{y} = 104.521 - 32.187(.44) + 27.688(-.5) - 9.354(.44)(-.5) + 24.729(+1)$$
$$\phantom{\hat{y} =} -6.563(.44)(+1)$$
$$= 100.414$$

(which approximately confirms our contour plot!).

Now our task is to see if we can hit 100.414 inches with the statapult pull back angle, hook, and peg set at those levels. We recommend 4 - 20 confirmation runs. (Rule of thumb: 4 runs if expensive, 20 runs if cheap.) After making six confirmation "shots," our data looked like:

A	B	C	y_1	y_2	y_3	y_4	y_5	y_6
0	−.5	+1	103	106	96	100	101.5	96

The average of the responses is:
$$\overline{CR} = \frac{103 + 106 + 96 + 100 + 101.5 + 96}{6} = 100.417$$

Does that confirm our prediction equation? Graphically, it looks like our prediction model does very well:

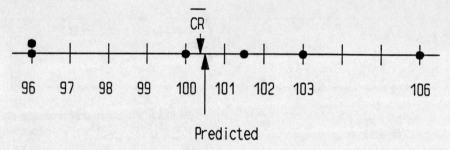

We will use a statistical confidence interval for the mean of our confirmation results to get a better feel for the accuracy of our model.

$$(\text{Interval}) = \overline{CR} \pm 3\left(\frac{S_{CR}}{\sqrt{n_{CR}}}\right) \quad [1]$$

In our example, $\overline{CR} = 100.417$
$S_{CR} = 3.95$

$$= 100.417 \pm \frac{3(3.95)}{\sqrt{6}}$$

$$= 100.417 \pm 4.844$$

$$= (95.57, 105.26)$$

If our predicted value for the given factor settings falls within this interval, we will consider our model good! In our statapult example, we are well within our confidence interval for the mean:

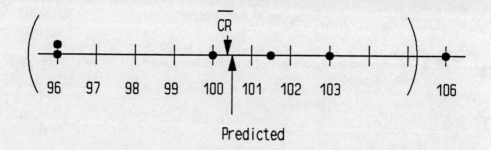

If not, we would start looking for the reason. If our predicted value is close to the upper or lower limits, we may make more confirmation runs. If that is not satisfactory, we should go back to Chapter 3 and begin digging into those eight reasons for poor experimental results.

Chapter 6 Bibliography

1. S. R. Schmidt and R. G. Launsby, *Understanding Industrial Designed Experiments (3rd edition)*. Colorado Springs, CO: Air Academy Press, 1991.

Case Study II-1:
Screening Experiment Using an L_{12}

A high technology computer company located in Vermont had purchased an RTC SMD-924 infrared oven for reflow of solder paste for surface mount devices. The ultimate objective of the experiment was to characterize the time/temperature profile of the oven. Initially, the company conducted a screening experiment to determine which factors had the greatest effect upon key responses.

Five responses were of importance to the engineers:

Table II.1 Factors

	Responses	Units of Measurement
R_1	Rate of temperature rise	°C/sec
R_2	Dwell from 150°-183°C	seconds
R_3	Maximum reflow temperature	°C
R_4	Time above 183°C	seconds
R_5	Time above 150°C	seconds

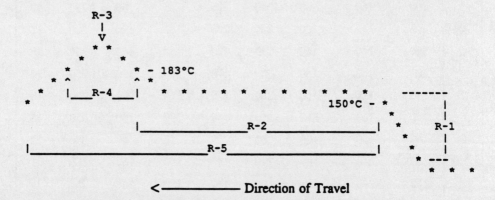

Figure II.1 Typical Temperature Profile (shown by Asterisk marks)

Factors and levels considered for the experiment were:

Table II.2 Factors and Levels

Factors	Low	High	Units
Air in zone 1 (az1)	100	160	scfh
Air in zone 2-4 (az24)	160	240	scfh
Air in zone 5 (az5)	100	160	scfh
Temp in zone 1 (t1)	270	300	°C
Temp in zone 2 (t2)	170	200	°C
Temp in zone 3 (t3)	160	180	°C
Temp in zone 4 (t4)	160	190	°C
Temp in zone 5 (t5)	250	300	°C
Belt Speed	18	22	ipm

An L_{12} orthogonal array was set up to screen out important variables for each response:

Table II.3 L_{12} Design

Run	Air Zone 1	Air Zone 2-4	Air Zone 5	Temp 1	Temp 2	Temp 3	Temp 4	Temp 5	Belt
1	100	160	100	270	170	160	160	250	18
2	100	160	100	270	170	180	190	300	22
3	100	160	160	300	200	160	160	250	22
4	100	240	100	300	200	160	190	300	18
5	100	240	160	270	200	180	160	300	18
6	100	240	160	300	170	180	190	250	22
7	160	160	160	300	170	160	190	300	18
8	160	160	160	270	200	180	190	250	18
9	160	160	100	300	200	180	160	300	22
10	160	240	160	270	170	160	160	300	22
11	160	240	100	300	170	180	160	250	18
12	160	240	100	270	200	160	190	250	22

Data obtained from the orthogonal array was:

Table II.4 Data Collected

Run	R1	R2	R3	R4	R5
1	2.25	0	175	0	145
2	2.50	85	197	40	160
3	2.75	0	175	0	215
4	3.00	228	205	60	318
5	2.75	203	202	50	285
6	2.50	105	183	0	160
7	2.75	95	205	55	185
8	2.00	190	192	45	270
9	2.60	160	190	35	225
10	2.00	25	182	0	70
11	2.00	0	179	0	215
12	2.00	0	182	0	170

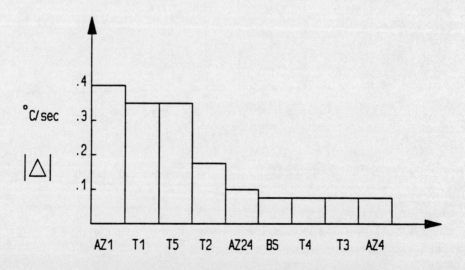

Figure II.2 Response (R_1) — Rate of Temperature Rise

For response R_1, air in zone 1, temperature in zone 1, and temperature in zone 5 appear to be the key factors.

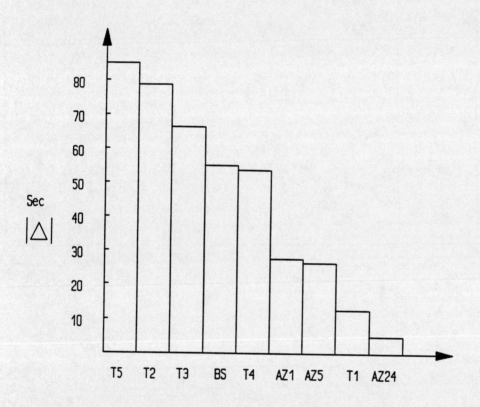

Figure II.3 Response (R_2) — Dwell Time from 150°-183°C

For response R_2, temperature in zone 5, temperature in zone 2, temperature in zone 3, belt speed, and temperature in zone 4 appear to be the key factors.

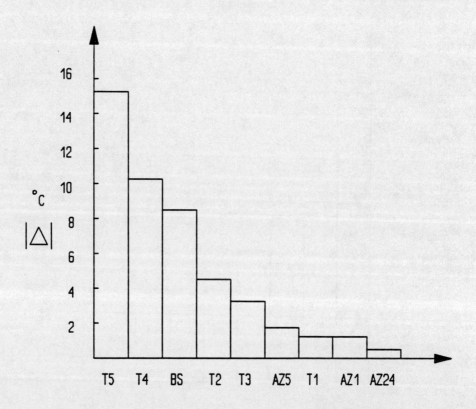

Figure II.4 Response (R_3) – Maximum Reflow Temperature

For response R_3, temperature in zone 5, temperature in zone 4, and belt speed appear to be the key factors.

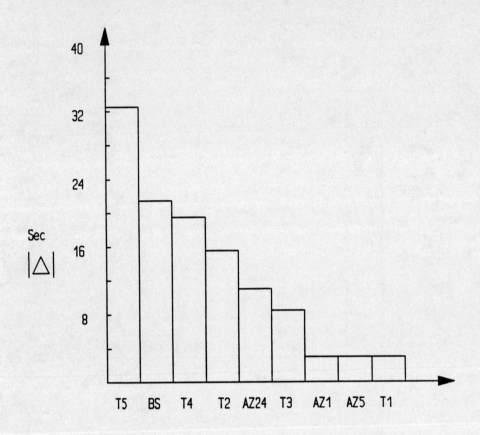

Figure II.5 Response (R_4) — Time Above 183°C

For response R_4, temperature in zone 5, belt speed, temperature in zone 4, and temperature in zone 2 appear to be most important.

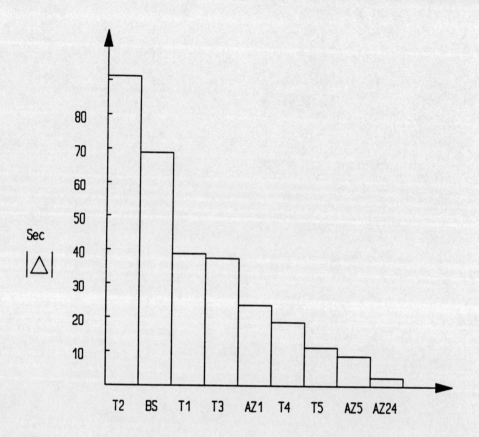

Figure II.6 Response (R_5) — Time Above 150°C

For response R_5, temperature in zone 2, belt speed, temperature in zone 1, and temperature in zone 3 appear most important.

Now, the company could concentrate on only the important factors in future testing or modeling activities.

Case Study II-2
Using Experimental Design to Design Tensile Test Samples
Submitted by Bradley Jones

One of the physical properties of a material is its tensile strength. Typically, strength is measured by physically pulling a sample of the material in two. The tensile strength is proportional to the applied force at the moment that the sample breaks. In practice, tensile testing is complicated by the fact that the tensile sample does not always break at the desired location. When this happens the test is useless. Therefore, it is desirable to design the shape of the test sample so that it is near certain that the sample will break at a specific point.

In this study, finite element analysis was used to simulate the stress on tensile samples of varying physical dimensions. A D-optimal statistically designed experiment defined the set of simulations performed. The result was a tensile test sample design that breaks in the center with high probability.

II.2.1 Background

Tensile test samples are shaped as in Figure II.7. To find the strength of the sample the test engineer fits the two holes over a pair of pins in a fixture. Hydraulic pressure then moves the pins apart applying stress to the sample. The pressure increases until the sample breaks, preferably in the thin region at the center of the sample.

Often, however, the stresses at the holes are great enough that the sample breaks at one of the holes instead of the center. When this occurs the sample is wasted.

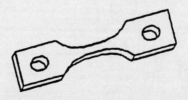

Figure II.7 Tension Specimen

Four design parameters were considered to affect the relative stress at the hole. These are illustrated in Figure II.8.

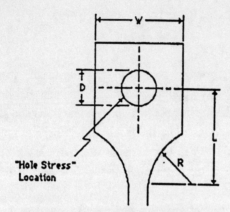

Figure II.8 Design Parameters

The purpose of this study was to optimize the design of the tensile test sample so that it was unlikely that the sample would break anywhere except the center.

II.2.2 Method

It was not feasible to actually make a number of tensile samples with varying head width, hole diameter, gage radius and hole location. Instead finite element analysis was used to simulate the stress contours on a theoretical part with given values of the above factors.

The design engineer had some prior experience with tensile sample design and chose reasonable ranges around his best current design. Factors and levels for the experiment were as follows:

Table II.5 Factors and Levels

Factor	Low	Middle	High	Units
Head Width (W)	12.7	15.9	19.1	mm
Hole Dia. (D)	5.0	6.5	8.0	mm
Gage Radius (R)	19.1	--	25.4	mm
Hole Loc. (L)	20.0	--	26.0	mm

Additionally, large 2-factor linear interactions were suspected between Head Width and Hole Diameter as well as Hole Diameter and Hole Location.

The result of the finite element analysis is a stress contour of the plot. The experimental response is the relative stress between the hole and the center of the tensile sample. If the stress at the hole and in the center are equal, then the relative stress is 100%. Values less than 100% imply that the hole stress is less than the center stress. The goal was to minimize this response. Using Catalyst (a BBN Software Products DOE package) software a 12 run D-optimal design was generated. The design and resultant data appear in Table II.6 below.

Table II.6 Optimal Design and Data Collected

Run	W	D	R	L	% Stress
1	12.7	5	19.1	20	110.849
2	19.1	8	19.1	20	94.534
3	12.7	5	25.4	26	108.166
4	19.1	5	25.4	20	91.654
5	12.7	8	25.4	20	179.270
6	12.7	8	19.1	26	163.487
7	19.1	5	19.1	26	77.551
8	19.1	8	25.4	26	81.038
9	15.9	6.5	19.1	20	98.908
10	15.9	8	25.4	26	100.074
11	15.9	5	25.4	26	85.912
12	12.7	6.5	25.4	26	124.234

II.2.3 Analysis

Analysis of the data using RS Discover [1] provided the following regression output table:

Table II.7 Regression Output (R^2 for the model is .903)

Term	Coeff.	Std. Error	T-value	Signif.
constant	91.340	4.796691		
W	-26.885	2.097292		
D	14.992	1.938947		
R	0.980	1.869853	0.52	.6365
L	-6.520	1.869853		
DW	-14.663	2.124010	-6.90	.0062
DL	-2.858	1.938947	-1.47	.2369
D^2	8.160	5.026326	1.62	.2030
W^2	13.535	4.234725	3.20	.0495

Refitting the regression with only the important terms (those with "significant" values less than .05 provided the following final regression table:

Table II.8 Final Regression Output (R^2 for the model is .9732)

Term	Coeff.	Std. Error	T-value	Signif.
constant	96.88	4.13		
W	−26.10	2.38		
D	14.42	2.23		
L	−5.84	2.08	−2.80	.03
DW	−14.66	2.49	−5.88	.001
W^2	15.37	4.74	3.24	.012

Taking the above equation and generating Response Surface Plots and Contour Plots provides the following graphics:

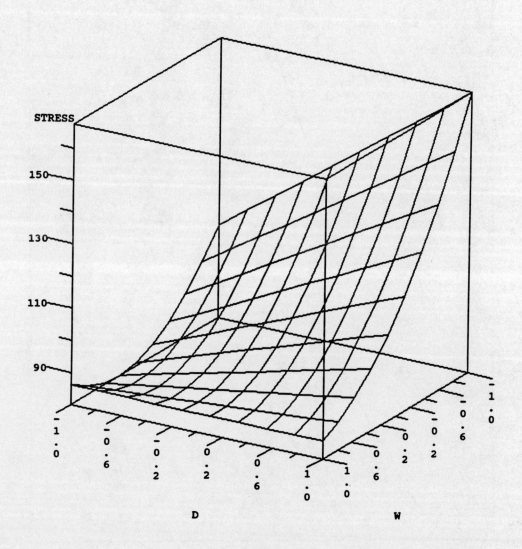

Figure II.9 Response Surface Plot of Stress
(Hole location set at the mid value and gage radius set at the mid value.)

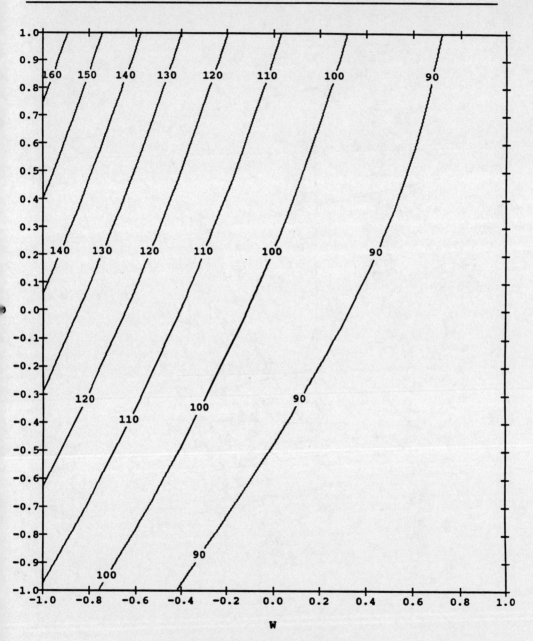

Figure II.10 Contour Plot of Stress
(Hole location and radius set at the mid value.)

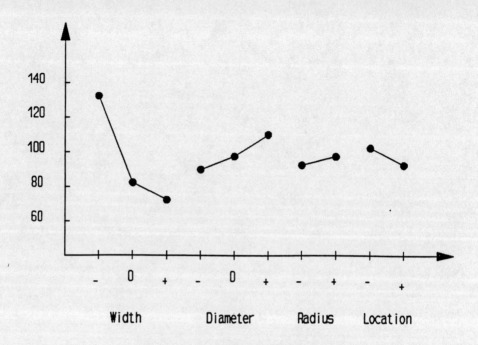

Figure II.11 Simple Graphical Analysis

Coupling the above simple graphical analysis with the Contour map and Response Surface Plot, the best settings for our factors are:

Width = 19.1 (high)
Diameter = 5.0 (low)
Gage Radius = does not seem to make a difference
Hole Location = 26.0 (high)

Confirmation tests and production runs indicated the test samples produced at the above settings produced test samples with no failures.

III. Robust Design

CHAPTER 7
INTRODUCTION TO ROBUST DESIGNS

7.1 What Is a Robust Design?

The word "robust" means insensitive. We want to be able to purchase low cost, state-of-the-art products which perform under varying conditions. Examples of typically robust consumer products include automobiles, automobile tires, television sets, cartridge disk (CD) players, telephones, etc. They all normally provide the required functions during varying usage conditions for the required life of the product. Examples of products which were not robust include the O-rings in the space shuttle *Challenger* (which became brittle in low temperatures), the carburetors on some older model automobiles (the chokes would not open automatically in cold weather), and the battery on our office Everex notebook computer (it stopped holding a charge for over 15 minutes after it was three weeks old).

Robust design specifically means:
1. Products perform intended functions without failure at varying customer usage conditions.
2. Allowance is made for:
   a. wide variation in customer usage.
   b. product deterioration (life).
   c. wide variation in subsystem/component parts.

Robust design recognizes that variability exists and is the enemy of high quality products and processes. The robust designer employs experimental design as a strategic weapon. It is accomplished by selecting the best levels for control factors such that the product performance is insensitive to those (noise) factors we either cannot or choose not to control.

7.2 Identifying Factors for a Robust Design

Since the objective of robust design is to make a product insensitive to noise, we must distinguish those factors we can control from those we cannot or choose not to control. You should recall from page 5 that we referred to those we won't control as noise factors. When discussing robust design it will also become important to identify those control factors that significantly affect the average response. These factors will be known as signal factors. All of these are laid out in the process model in Figure 7.1.

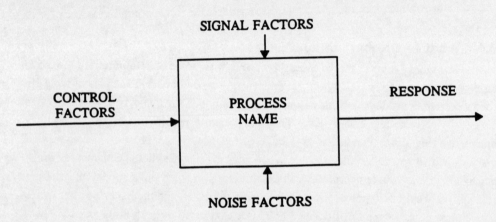

Figure 7.1 Layout of a Robust Design

7.2.1 Control Factors

Based on how they affect the mean and variation of the response, there are four types of control factors as shown in Figure 7.2 on the following page. Since Factor A only affects the response average it is the signal factor we mentioned above. We will discuss signal factors in Chapter 9. Factor A is a very common type of factor in most processes. Factor B is a rare (but wonderful) factor. As displayed in the graphic, the average response is the same regardless of whether factor B is set at level 1 or 2. The variability, however, is less when B is set at level 1. Factor C is relatively common in that it affects both the average and variation. In fact, the variability typically increases when the average increases. Factor D, on the other hand, does nothing. Factors like D are also useful to know about for cost purposes. If one level is lower cost than the other we will select the low cost level for Factor D.

i) affect location

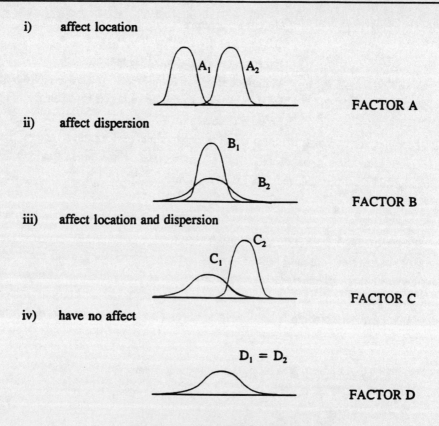

ii) affect dispersion

iii) affect location and dispersion

iv) have no affect

Figure 7.2 Four Types of Factors

7.2.2 Noise Factors

There are three types of noise factors we must be cognizant of:
- EXTERNAL   Customer imposed during usage.
- INTERNAL   Variations resulting from differences in raw material or variations resulting from our manufacturing process.
- DEGRADATION   Life requirements.

A noise factor is any factor which is known (or believed) to affect our response, but we either cannot or choose not to control. Examples of noise factors relating to a company designing and manufacturing computer disk drives might be:

- **EXTERNAL**	Usage temperature
Usage humidity
Interface (work station, PC, notebook)
- **INTERNAL**	Manufacturing variation (heads, disks, circuit boards)
- **DEGRADATION**	Product needs to provide required function for life of product (five years).

The challenge is for the disk drive to provide outstanding performance (access time, transfer rate, error free storage and retrieval) regardless of the noise environment. Additionally this must be achieved in a low-cost, timely manner.

7.3 When Should You Do Robust Design?

Many companies wait until release to manufacturing to attempt to make the product robust (the "over-the-wall" engineering approach). This typically leads to excessive costs and missed target dates for shipments. As shown in Figure 7.3, our ability to influence a development project outcome is high early in development (diagonally shaded area).

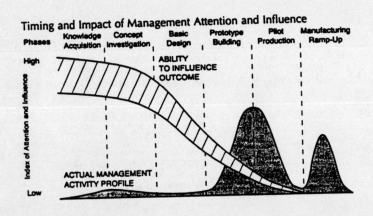

Figure 7.3 Timing and Impact of Management Attention and Influence [1]

Unfortunately, our actual activity level tends to become great late in the project, when the effort is in trouble. The focus of experimental design and robust design techniques must be on the early stages of the development effort.

To introduce robust design as early as possible, Hewlett Packard Corporation has implemented the overall development strategy shown in Figure 7.4 [2]. This strategy stresses the separation of technology invention and its application in new product development. The "Pizza Bin of Proven Technologies" includes only those technologies which have been modeled, characterized, and proven to be robust. Only these technologies (components, subsystems) are integrated with other proven subsystems to create a system level concept for the new development project. Robust design techniques will be applied at the system level as well. Finally, pilot production and manufacturing ramp-up will proceed much more smoothly and timely than if robust design techniques had not been applied at the technology or development project system level.

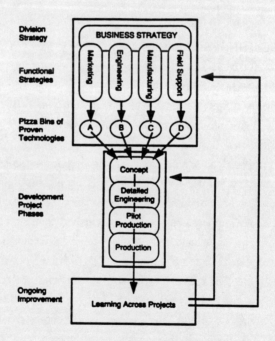

Figure 7.4 Overall Development Strategy at Hewlett Packard

Continuing, we can think of the development of a new product as consisting of three phases [3]:

CONCEPT DESIGN: The engineering invention stage. It involves the early development effort. Modeling types of experiments as well as robust design of experiments are to be done at this phase.

PARAMETER DESIGN: Involves taking the various technologies, components, and subsystems and integrating them into a low-cost, robust system. Stated in a slightly different manner: In parameter design we attempt to select the best levels of our control factors such that we get excellent system performance regardless of the levels of our noise factors. If this can be accomplished in a low-cost manner, we have probably attained a competitive advantage. If we are unable to make the system totally robust at this phase, we must go to the third stage, tolerance design, to make the system robust.

TOLERANCE DESIGN: In this stage, we typically have to add cost or tighten tolerances to make the system robust.

Let's address a simple example which will help clarify the set-up of a robust design [4] and how it fits into the three phases of the development process.

EXAMPLE 7.1 Suppose you are an engineer working for a company which produces caramel candies. At room temperature, the caramels are just the right hardness to delight the palettes of your toughest customers. Unfortunately, the hardness of the product is very sensitive to usage temperature. Suppose you had some of the product in your backpack while skiing on a blustery, cold, January day in Breckenridge, Colorado. After an hour of skiing you decide to sample the caramels. Imagine your displeasure when you realize the caramels are like rocks. Of course, the opposite problem occurs when the product is exposed to heat. Imagine toting a few caramels to the beach on a warm summer day. In only a few minutes the caramels will become a kind of caramel soup.

Several ways this sensitivity issue could be addressed are:
1. Reformulate the product (time consuming, additional cost).
2. Place an environmental chamber around each caramel (costly).
3. Put a warning label on each caramel suggesting consumers only consume the product at room temperature (may limit market penetration).
4. Formulate several different products for different climates (costly;

marketing and distribution nightmare).

The first idea suggests we return to **CONCEPT DESIGN**. The remaining ideas are **TOLERANCE DESIGNS**. We have completed ignored the possibility of **PARAMETER DESIGN**. Sadly, this is the design process we've encountered in many companies.

Let's suppose that this is a somewhat progressive company and that they will make their product robust early in the development process.

CONCEPT DESIGN: The company believes that "Grandma's recipe" is sound. They do not believe they should put their efforts here.

PARAMETER DESIGN: Instead of jumping right to tolerance design, suppose the company attempted a robust design at this phase. After conducting a functional analysis based upon customer needs, the following process model was obtained.

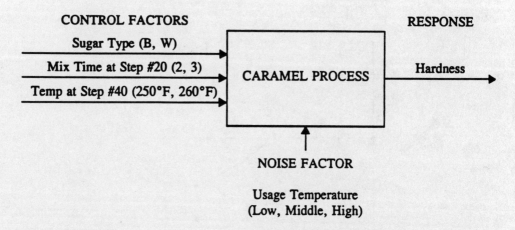

Three control factors were used. One noise factor (usage temperature) was identified. The control factors were assigned to an L_4 orthogonal array (this is typically called the inner array). The noise factors will also be assigned to an orthogonal array (the outer array). In our example the outer array is very simple in that it consists of one factor at three levels. The test matrix is shown in Table 7.1.

Table 7.1 Inner and Outer Array for Caramel Candy Problem

Run	Sugar Type	Mix Time	Temp	Usage Temperature		
				Low	Middle	High
1	B	2	250°F			
2	B	2	260°F			
3	B	3	250°F			
4	B	3	260°F			
5	W	2	250°F			
6	W	2	260°F			
7	W	3	250°F			
8	W	3	260°F			

Using this array, we will test each recipe (or run) at each usage temperature setting. Our hope is to find a run where the hardness is on target with little variability. Before we look at the results, let's discuss the "before" and "target" conditions.

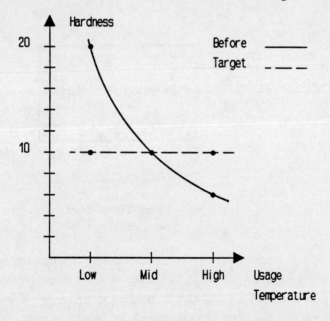

Figure 7.5 Hardness vs Temperature

As displayed in Figure 7.5, suppose the target hardness is 10 units. As discussed, the "before" condition is characterized by a great amount of variability in hardness as usage temperature changes.

Table 7.2 Inner and Outer Array for Caramel Candy Problem

				Usage Temperature		
Run	Sugar Type	Mix Time	Temp	Low	Middle	High
1	B	2	250°F		X	
2	B	2	260°F			
3	B	3	250°F			
4	B	3	260°F			
5	W	2	250°F			
6	W	2	260°F			
7	W	3	250°F			
8	W	3	260°F			O

In Table 7.2, let's discuss the conditions under which hardness will be measured in the cell with "X" in it. We will use sugar type "B", mix time = 2, and cook temperature at Step #40 at 250°F. The caramels will then be packaged. A portion of this run will then be conditioned to a "middle" temperature. Once the product has been properly conditioned, the hardness will be measured. In Table 7.2, where the "O" is shown, the caramels will be produced using sugar type "W", mix time = 3, and cook temperature = 260°F. After the product has been packaged a portion will be conditioned to a "high" usage temperature. Once the caramels have been properly conditioned, the hardness will be measured and recorded in the cell with "O" in it. In order to complete the robust design, hardness values will need to be recorded in all 24 cells.

Suppose all 24 trials were conducted per the orthogonal array and the data obtained was as shown in Table 7.3. Notice that run #5 is the winning combination. Not only is the average hardness on target, but there is little variability in face of noise. If the variability shown in combination #5 is acceptable to their customer, they are ready to CONFIRM their results. If it is not, the company may be forced into some tolerance designing.

Table 7.3 Completed Caramel Candy Experiment

Run	Sugar Type	Mix Time	Temp	Usage Temperature		
				Low	Middle	High
1	B	2	250°F	10	10	0
2	B	2	260°F	19	30	1
3	B	3	250°F	17	14	2
4	B	3	260°F	22	15	3
5	W	2	250°F	11	10	9
6	W	2	260°F	15	12	3
7	W	3	250°F	18	14	7
8	W	3	260°F	20	12	2

TOLERANCE DESIGN: Continuing with our caramel candy example, let's change the scenario. Suppose the customer requirement is 10 ± .3 for hardness, regardless of the usage temperature. Run #5 has gotten us close to our requirement, but not totally there. In this case it appears we have obtained great improvement (but not the required improvement) from parameter design. This will be the case many times in actual practice of the techniques. Since we have failed to make the product totally robust through parameter design, we must go to the third stage, tolerance design. Our approach now would be to address the robust problem by attempting one or more of the remedial measures previously discussed.

In the next chapter, we will learn about several approaches to analysis. Fundamentally, however, we want to do two things in a robust design:
1. Reduce the variability in the face of noise.
2. Place our process average on target.

Unfortunately, in many companies we neglect to integrate fully modeled robust technologies into our systems. Commonly, we wait until pilot production and manufacturing ramp-up to make our products robust. The use of experimental design techniques early-on can be (and is) a source of competitive advantage for many leading world class companies.

Chapter 7 Bibliography

1. R. H. Hayes, S.C. Wheelwright, and K. B. Clark, *Dynamic Manufacturing*. New York: The Free Press, 1988, p. 279.

2. S. C. Wheelwright and K. B. Clark, *Revolutionizing Product Development*. New York: Macmillan, Inc., 1992, p. 40.

3. S. R. Schmidt and R. G. Launsby, *Understanding Industrial Designed Experiments (3rd edition)*. Colorado Springs, CO: Air Academy Press, 1991.

4. American Supplier Institute, Two-week Taguchi Techniques Seminar, Dearborn Michigan, Fall 1985.

CHAPTER 8
APPLYING THE BOX AND BUBBLE TO ROBUST DESIGNS

The steps involved in conducting a robust design are exactly the same as we discussed in Chapters 1–6. We must modify our thinking, however, in some of them.

8.1 Planning

When planning, it is important to identify both control and noise factors. We spent a lot of time in Section 7.2 discussing both. It is also just as important to select experimental levels for your noise factors as it is for your control factors. Perhaps it would be most useful to concentrate on a couple of examples:

EXAMPLE 8.1 Imagine we are a custom injection molding operation in Boulder, Colorado. One product we make in substantial volume is roller blade rollers.

Figure 8.1 Roller Blade Rollers

Recently our customer has complained that they are obtaining rollers with different diameters. (Their customers do not find this to be a redeeming feature for the product to have.) Figure 8.2 provides a diagram of the bottom half of our die.

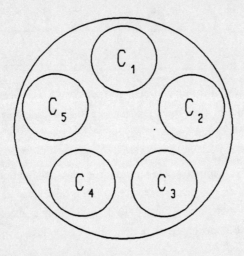

Figure 8.2 Roller Blade Die

Notice the die has five cavities (each is uniquely identified). Initially, we suspect the cavities may have been machined to different diameters. Upon review, however, we find they are all the same. Further analysis suggests that because some cavities are closer to cooling lines than others, the polymer in each cavity sees a different time/temperature profile. This may provide each molded part with unique shrinkage characteristics. After brainstorming the process, we determine there are three control factors of interest and one noise factor. The factors are:

CONTROL FACTORS			NOISE FACTORS
Factor	Low(−)	High(+)	
Preht Temp	150°F	160°F	Cavity # (C_1, C_2, C_3, C_4, C_5)
Zone Temp 1	270°F	280°F	
Zone Temp 2	280°F	285°F	

EXAMPLE 8.2 Let us consider an example from the copier business. Suppose you are a design engineer doing system design for a new copier. The product is essentially a leveraged product with a new paper feed mechanism. Suppose the control factors to be included in the experiment are:

Factor	Low(−)	High(+)
Vacuum required to pull one sheet from stack	1	2
Paper Stack Location "A"	1	2
Paper Stack Location "B"	1	2

The noise factors (and anticipated levels) selected were:

> Paper Weight (12, 20, 32, 120)
> Paper Size (5x7, 8.5x11, 11x17)
> Paper Skew (1, 2, 3)
> Manufacturer of Paper (X, K, BC, JR)
> Copier Operating Humidity (L, M, H)
> Copier Operating Temperature (L, M, H)

Note that the noise factors we have introduced with these two examples have multiple levels. In Example 8.2, we also have many factors. Multiple factors (with) many levels can dramatically increase the complexity of the experiment. After we see the size of the design matrix required for our experiment (Section 8.2), we may have to return to the planning phase to try to reduce our levels.

8.2 Selecting an Orthogonal Array

In selecting an orthogonal array we will want to assign control factors to an inner orthogonal array and noise factors to an outer orthogonal array. We have been using an inner array all along. The inner array is the location of our control factor settings. The outer array is the matrix of noise factor settings normally placed on its side above the area we reserve for our response values.

8-4 Straight Talk on Designing Experiments

```
                    ┌─────────────────────────────┐
                    │ N │                         │
                    │ o │                         │
                    │ i │   Matrix of Noise       │
                    │ s │   Factor Settings       │
                    │ e │   (Outer Array)         │
                    │ F │                         │
                    │ a │                         │
                    │ c │                         │
                    │ t │                         │
                    │ o │                         │
                    │ r │                         │
                    │ s │                         │
                    ├───┼────────────┬────────────┤
                    │   │ Control Factors │       │
                    │Run├────┬───┬───┬───┤        │
                    │   │    │   │   │   │       │
                    ├───┴────┴───┴───┴───┼────────┤
                    │                    │        │
                    │  Matrix of Control │ Matrix │
                    │  Factor Settings   │of Resp.│
                    │  (Inner Array)     │ Values │
                    │                    │        │
                    └────────────────────┴────────┘
```

For example, consider two control factors and two noise factors at two levels. The full design matrix using full factorials for the inner and outer arrays would look like:

| | Noise Factors | Factor 1 | − | + | − | + |
		Factor 2	−	−	+	+
	Control Factors					
Run	Factor 1	Factor 2	Response			
1	−	−				
2	−	+				
3	+	−				
4	+	+				

Returning to Example 8.1, the design matrix would look like:

Run	Noise Factor			Cavity				
	Control Factors			1	2	3	4	5
	Preht	ZT 1	ZT 2	Cavity Diameter				
1	150°F	270°F	280°F					
2	150°F	270°F	285°F					
3	150°F	280°F	280°F					
4	150°F	280°F	285°F					
5	160°F	270°F	280°F					
6	160°F	270°F	285°F					
7	160°F	280°F	280°F					
8	160°F	280°F	285°F					

If we only produce one replicate for each cavity, we must make 40 parts. In some experiments that may be formidable. In this case, however, the parts are made quickly. The time required is in allowing the process to stabilize between runs, allowing the parts produced to cool a sufficient amount of time, and actually measuring the diameters of the parts. As you can imagine, with more than a few control and noise factors, the size of the array for robust design can become very large.

Now, recall Example 8.2. We listed three control factors at 2 levels, four noise factors at three levels, and two noise factors at four levels. If we were to conduct a full factorial inner array (8 runs) and a full factorial outer array (4x3x3x4x3x3 = 1296), the number of cells in our robust design would be 8x1296 = 10,368. Of course, other design options are available (fractional factorial, D-optimal, etc.) to help us limit the number of runs. You should be aware that the number of cells can quickly get out of hand with this type of approach. Because of this, we will need to address ways to limit the size of the array. Some possible options are:

1. Conduct a screening design on the control factors with noise factors held at some nominal level so as to determine the few important control factors.
2. If you know best case and worst case noise conditions, run only the best and worst case noise for each combination. If we get outstanding results at each of these conditions (and we can assume all else in between provides acceptable results) we can conclude a robust setting has been determined.
3. Run a screening design on the noise factors with the control factors held at nominal conditions to determine the extreme conditions. Identify a compound noise factor at two levels where the low level drives the noise to the low side and high level drives the noise to the high side.

As an example of option 3 above, suppose we decide to conduct a screening design on just the noise factors in our process with all control factors held constant at nominal values.

NOISE FACTORS	LOW	HIGH
A	1	2
B	1	2
C	1	2

	A	B	C
Avg 1	55	32	48
Avg 2	20	43	27
Δ	-35	$+11$	-19

COMPOUND NOISE (LOW), $N_1 = A_2B_1C_2$
COMPOUND NOISE (HIGH), $N_2 = A_1B_2C_1$

Before leaving the setup of a robust design, we need to make several summary comments. In order to find a robust combination, one or both of the following needs to take place:
1. An interaction between a control and noise factor.
2. Non-linearity between a control factor and the response.

If you use the inner and outer array arrangement you will be able to determine the most likely interactions between control and noise factors. In order to take advantage of (2), a three-level inner array is useful. Taguchi's tabled orthogonal arrays (L_9, L_{18}, L_{27}) as well as classical designs such as the central composite or even D-optimal designs can be used.

8.3 Conducting the Experiment

Follow the guidelines set forth in Chapter 3.

8.4 Analyzing Your Results

Analysis of a robust design does not have to be complex. Remember that we are attempting to select settings for the control factors such that the process is on target with minimum variation regardless of the settings for the noise factors. Two approaches will be discussed:

1. The "blended" approach
2. The Taguchi approach

The merits of both approaches have been debated vigorously during the last ten years. For details of this discussion read [1] and [2]. Our intent is to move beyond the debate and suggest that it is probably useful to analyze the data with both approaches. If both approaches provide the same best levels for the important factors (which should occur frequently) this is good. On the other hand, if the analysis suggests two different best conditions, simply run confirmations on both proposed solutions. The one which most closely matches the objective is the analysis method to choose.

8.4.1 The Blended Approach

First we will discuss the "blended" approach to the analysis of a robust design. The steps to be employed are:

1. For each row of the data, calculate Ln (log base e) of s (standard deviation).
2. Conduct simple analysis of the Ln s statistic. Rule of Thumb: If the $|\Delta/2|$ for Ln s ≥ 0.5, then you may have an important variance reduction factor [1].

3. Conduct regular analysis of the $\bar{y}$ column to find factors which shift the average.

To demonstrate the use of these steps, we will analyze a simple example using the statapult from Section I.

EXAMPLE 8.3 On the third day of our seminar we give each student group three different types of balls and ask them to complete a robust design. We ask them to find those factors that have a significant effect on the variation of downrange distance. Then we ask each group to hit a target distance that we select. The group that hits the closest to the target with all three balls is the winner. Because time is short we often limit the experiment to two control factors. After brainstorming the statapult, one group decided to use the following factors and responses (on some of our statapults, we have a protractor attached to measure pull back angle):

Control			Noise			
Factor	Levels		Factors	Levels		
	Low(−)	High(+)		Low	Middle	High
Pull Back (A)	70°	30°	Ball Type	Styro (S)	Smash (SM)	Golf (G)
Cup Position (C)	1	2				

Their experiment yielded the following results:

	Noise Factors	Ball Type	S	SM	G		
	Control Factors						
Run	A	C	Response			$\bar{y}$	s
1	70°	1	26	29	28	27.7	1.53
2	70°	2	36	38	44	39.3	4.16
3	30°	1	80	101	112	97.7	16.26
4	30°	2	97	122	144	121	23.5

Following the steps on the previous pages, the results of our analysis are:

STEP 1: Calculate ln s for each row of data.

Run	y	s	Ln s
1	27.7	1.53	.425
2	39.3	4.16	1.43
3	97.7	16.3	2.79
4	121	23.5	3.16

STEP 2: Analysis of ln s

Run	A	C	AxC
1	70°	1	+
2	70°	2	−
3	30°	1	−
4	30°	2	+
Avg Low	.928	1.60	2.10
Avg High	2.97	2.29	1.79
Δ/2	1.02	.34	−.15

The data suggests factor A (Pull back angle) may be an important variance reduction factor. The best setting for pull back angle appears to be the low setting (70°).

STEP 3: Analysis of the Average

Run	A	C	AxC
1	70°	1	+
2	70°	2	−
3	30°	1	−
4	30°	2	+
Avg Low	33.5	62.7	68.5
Avg High	109	80.2	74.3
Δ	75.9	17.5	5.8
Δ/2	37.9	8.8	2.9
F_c	82.4	4.4	.5

Recall that the "F" statistic for a two-level design is $F = \dfrac{MSB}{MSE}$ (see Chapter 4). F > 6.0 for a factor indicates that it significantly effects the response. F between 4 and 6 is questionable. The data suggests factors A & C may play a significant role in shifting the average (relatively large F_C). The interaction effect does not appear to be significant.

8.4.2 The Taguchi Approach
The steps to be employed are:

1. Based upon the OBJECTIVE, select the proper signal-to-noise (S/N) ratio. The S/N ratio is NOT that used in engineering. It is a concurrent statistic used to combine the mean and the standard deviation. Dr. Taguchi gave it the name S/N to encourage its use among his colleagues at the Electrical Communications Laboratory of the Nippon Telephone and Telegraph Company [3].

2. Calculate the selected S/N for each row of data.

3. Conduct simple analysis of the S/N statistic.
   RULE OF THUMB: If the $|\Delta|$ for S/N ≥ 2.0, you may have an important factor. For important factors, the level with the greatest value is the best level.

4. For signal-to-noise nominal is best (S/N_N), conduct regular analysis of the $\overline{y}$ column to find factors which shift the average. Use one or more of these factors to target the response.

Although Dr. Taguchi has developed over 70 S/Ns, most are proprietary [3]. However, the following are not, and will form the core of our discussion:

- **Smaller is Better** (Implies that we are trying to minimize a response while minimizing the variance.)

$$S/N_S = -10 \log_{10} \frac{1}{n} \sum_{1}^{n} \left(Y_i^2\right)$$

- **Nominal is Better** (Implies that we are attempting to place our response at some nominal value while minimizing the variance.)

$$S/N_N = 10 \log_{10} \frac{1}{n} \left(\frac{S_m - V_e}{V_e} \right)$$

where $S_m = \frac{(\Sigma Y_i)^2}{n}$ and $V_e = \frac{\Sigma Y_i^2 - \frac{(\Sigma Y_i)^2}{n}}{n - 1}$

- **Larger is Better** (Implies that we will maximize our response while minimizing the variance.)

$$S/N_L = -10 \log_{10} \frac{1}{n} \sum_{i}^{n} \left(\frac{1}{Y_i^2} \right)$$

- **Dynamic S/N Ratio** (This will be discussed in Chapter 9.)

$$S/N_D = +10 \log_{10} \frac{\beta^2}{MSE}$$

Once again, we will use our statapult to demonstrate these steps.

STEP 1: Since our objective is to hit a target value, we decided to use the S/N_N.

STEP 2: Using the statapult data from section 8.4.1 we have computed the S/N_N ratio for each run:

8-12 Straight Talk on Designing Experiments

	FACTORS		BALL TYPE					
Run	A	C	S	SM	G	$\bar{y}$	s	S/N_N
1	70°	1	26	29	28	27.7	1.53	25.16
2	70°	2	36	38	44	39.3	4.16	19.51
3	30°	1	80	101	112	97.7	16.26	15.57
4	30°	2	97	122	144	121.0	23.52	14.23

STEP 3: Simple Analysis of the S/N_N looks like:

Run	A	C	AxC
Avg Low	22.33	20.37	17.54
Avg High	14.90	16.87	19.69
Δ	−7.43	−3.5	2.15

The Δ's for A and C appear to be great. The interaction effect appears to be of marginal importance. Since the guidelines for selecting which effects are to be considered large are very general [4], let us assume only A and C have important effects. This seems to concur with the blended approach results in Section 8.4.1.

STEP 4: Regular analysis of y:

Run	A	C	AxC
Avg Low	33.5	62.7	68.5
Avg High	109.4	80.2	74.3
Δ	75.9	17.5	5.8

As we previously found in Section 8.4.1, both A and C appear to have important effects on the average.

8.5 Put Your Process on Target and Confirm

8.5.1 The Blended Approach

Since we would like to set factor A (pull angle) at the low setting for reduced variation, we will attempt to use cup position to move the distance to a target value of 50 inches. Unfortunately, factor C (cup position) is a qualitative factor. Only position 1 or 2 can be chosen. The contour plot given in Figure 8.3 indicates two possible solutions for the problem. Either cup position #1 can be chosen with a pull back angle of $\approx 57.2°$ or cup position #2 can be chosen with a pull back angle of $\approx 65.0°$. Because analysis of Ln s indicated a pull back angle of 70° provided a smaller amount of variability in the distance, we would probably select the setting with a pull back angle at 65° with cup position #2. The next step is to perform confirmation tests. In this case, cup position would be placed at position #2, pull back angle would be positioned at 65° and we would toss all three balls (a total of 4 - 20 times) to see if the experiment confirmed.

8.5.2 The Taguchi Approach

Since only factor A is adjustable on a continuous basis, we will set factor C at the best level (low) and use factor A to adjust the average to the target value of 50 inches. For either approach, the only work remaining is to complete 4 - 20 confirmation runs at the target settings.

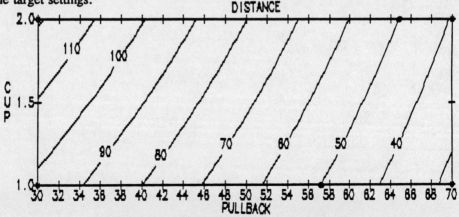

Figure 8.3 Contour Plot of Statapult Model [5]

Chapter 8 Bibliography

1. Schmidt, S. R. and Launsby, R. G., *Understanding Industrial Designed Experiments (3rd edition)*. Colorado Springs, CO: Air Academy Press, 1991.

2. Bendell, A.; Disney, J.; and Pridmore, W. A., *Taguchi Methods, Applications in World Industry*. UK: IFS Publications, 1989.

3. Barker, T. B., *Engineering Quality by Design*. New York: ASQC Quality Press, 1990.

4. Taguchi, G., *Taguchi Methods, Research, and Development*. Dearborn, MI: American Supplier Institute, 1992.

5. *RS Discover*, BBN Software Products Corporation, Cambridge, MA.

CHAPTER 9
PRODUCTS WITH DYNAMIC CHARACTERISTICS

Many products exhibit what can be called dynamic characteristics. For example, when turning an automobile, the driver determines the turning radius by turning the steering wheel. The greater the degree of turn of the wheel, the sharper the turn. On photocopying machines, the person making the copy can influence the darkness of the resulting copy by adjusting the shading button. Skillful golfers can adjust the flight distance and height of their shot predominantly by the selection of the club. Systems with dynamic characteristics can be described by the following flow chart:

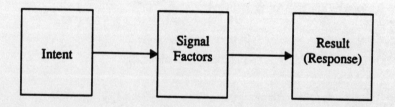

For example, suppose you wish to make a right turn (90°) from a complete stop. Your INTENT or intention is to make a right turn. The steering wheel is the means provided for expressing this intent. The SIGNAL FACTOR is the angle you turn the wheel. The RESULT or response is that the vehicle turns at the required angle. In a second example, perhaps you wish to toss a small rubber ball with a catapult (see Chapter 1 for a picture and discussion). Your INTENT is to toss the ball 100 inches. The angle the arm is pulled back is the means of expressing this intent. The SIGNAL FACTOR (arm) is positioned at an angle of 150° (measured on the attached protractor) and then released. The RESULT or response is the ball hits the base of a target positioned 100 inches from the nose of the catapult.

Continuing with the concept of dynamic characteristics we can modify the process model first introduced to incorporate Taguchi's breakdown of inputs (factors) in the model.

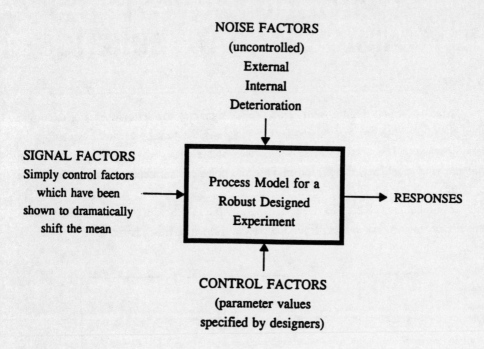

With robust design we will attempt (through the use of orthogonal arrays) to find the settings for our control and signal factors so that our products will perform well regardless of the influence of uncontrolled or noise factors.

In Figure 7.2 on page 7-3, we originally identified four types of factors which could be identified through designed experiments. Factor A which affects the response location (but not variation) is, using the Taguchi methodology, a signal factor. Signal factors were originally introduced by Taguchi as those which the customer can use to adjust the response. Recent Taguchi teachings (after 1991) suggest any control factor which has a great impact on the response mean, but no effect on the variation, can be identified as a signal factor. When we look at Taguchi's recent enhancements to the analysis of robust designs using dynamic characteristics we will better appreciate why Taguchi makes this distinction.

9.1 Planning

Except for identifying signal factors, this phase will be the same as we discussed in Chapter 8. For illustration, let's consider our statapult. We have determined that the pull back angle significantly alters the downrange distance and will be our signal factor. The remaining factors are:

SIGNAL FACTOR: A = Pull back angle (35°, 70°)
CONTROL FACTORS: S = Stop location (1, 2)
                   C = Cup location (1, 2)
NOISE FACTOR: Ball type (R = rubber, W = whiffle, S = sponge)

9.2 Selecting an Orthogonal Array

Figure 9.1 tells you everything you need to know. Once again we will look at the blended approach and the Taguchi approach.

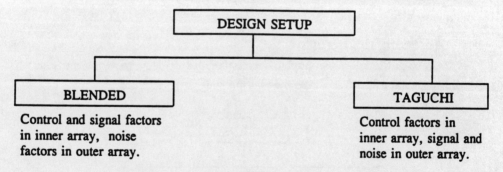

Figure 9.1 Design Setup for a Robust Design

9.2.1 The Blended Approach

There's nothing new to add here. The beauty of this approach is its consistency.

9.2.2 The Taguchi Approach

Dr. Taguchi places both signal and noise factors in the outer array. For the statapult, that would look like:

Run	Signal Factor		35°			70°		
	Noise Factor		R	W	S	R	W	S
	Control Factors							
	Stop	Cup						
1	1	1						
2	1	2						
3	2	1						
4	2	2						

9.3 Conducting Your Experiment

Follow the guidelines in Chapter 3. For our current statapult example, two repetitions were made of each run.

9.4 Analyzing the Results

Although the following graphic depicts the analysis flow as very similar to what we did in Chapter 8, this is where the two methods diverge in complexity.

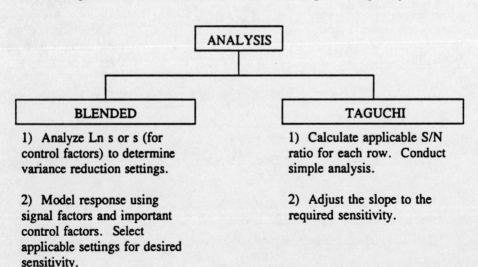

9.4.1 The Blended Approach

This should be very familiar to you by now. Using more statapult data with two replicates of each run:

	Signal Factor	Noise Factor	Ball Type	R		W		S		y	s	ln s
		Control Factors										
Run	Angle	Stop	Cup									
1	35	1	1	53	53	52	52	57	53			
2	35	1	2	35	34	36	33	34	33			
3	35	2	1	45	46	55	48	57	52			
4	35	2	2	35	30	36	40	38	40			
5	70	1	1	108	108	109	105	107	105			
6	70	1	2	85	84	87	86	86	88			
7	70	2	1	95	96	105	98	107	99			
8	70	2	2	85	86	82	80	88	81			

STEP 1:   Ln S AND S ANALYSIS

                 Stop is possible variance reduction.
                 Set factor at the low setting.

STEP 2:   MODEL DISTANCE

$$\hat{y} = 68.71 + 25.4(A) - 8.6(C) - 1.3(A)(S) - 1.0$$

                 Use the prediction equation to hit target.

In Step 1 of this approach, we analyze only the control factors. Use Ln s (or s) to find possible variance reduction factors. Our analysis suggests setting stop at level 1 may provide us with this information. In Step 2, using both the angle and cup (we will set s = 1 for reduced variation) a prediction model for distance is generated.

9.4.2 The Taguchi Approach

Recent teachings from Taguchi suggest nearly any problem which previously was evaluated as a static problem can be addressed using a dynamic signal-to-noise ratio. The general form of the dynamic signal-to-noise ratio was introduced in Chapter 8 as: $S/N_D = 10 \log_{10} \frac{\beta^2}{MSE}$. MSE and β will be calculated differently based upon the response type, factor type, and range of interest in the response. The following flow chart helps with the breakdown of the competing cases.

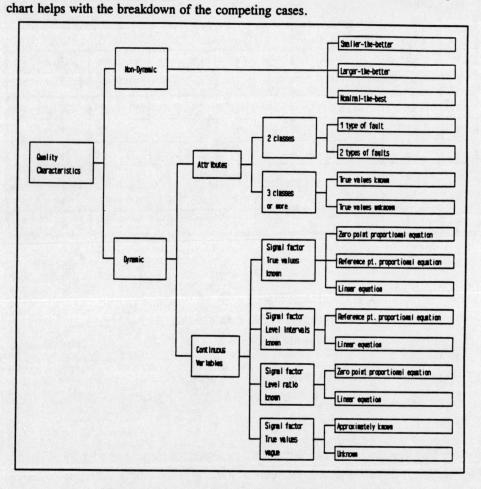

Figure 9.2 Quality Characteristics (S/N Ratios)[1]

The previous figure suggests the selection of the proper S/N ratio is formidable. Because of Taguchi's emphasis on continuous responses, however, only a few cases are emphasized. Because of this, we will focus on the "signal factor true values known" cases:

1. "Zero point proportional" — Assumes the response is zero when the signal factor setting is zero. In this case a model of the form $\hat{y} = \beta x$ can be fit to the row data to find β and MSE.

2. "Reference point proportional equation" — Again assume the response is zero when the signal factor is set at zero. In this case, however, we would recode the response values (typically by subtracting the mean from each point) so as to focus our attention over the immediate range of our response values. An equation of the form $\hat{y} = \beta x$ is then fit to the data.

3. "Linear equation" — Used when it is known the response value $\neq 0$ when the signal factor is set at zero. An equation of the form $\hat{y} = \text{constant} + \beta x$ is fit to the row data.

Using the statapult data from Section 9.4.1, we will apply the Taguchi approach. To compute MSE and β we will use the "zero point proportional" case since the in-flight distance is zero as long as the arm is never pulled back. With both noise and signal factors in the outer array, the first run looks like:

Run	Stop	Cup	35°			70°			β	MSE	S/N_D
			R	W	S	R	W	S			
1	1	1	53	52	57	108	109	107			
			50	52	53	108	105	105			

To make you feel a little more comfortable with the S/N_D, we will guide you through the β and MSE computation for run 1. Plotting the response vs. the signal factor (Pull back angle), we have:

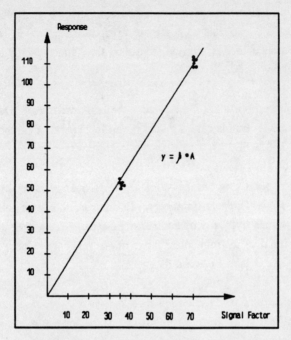

Figure 9.3 Pull Back Angle vs. Distance

Using the "zero point proportional" case, we expect our prediction line to go through the origin. Using MYSTAT to perform the simple linear regression, the output table looks like:

Table 9.1 Regression Output for Run #1 S/N_D Parameters

```
MODEL CONTAINS NO CONSTANT
   Dep Var: Y      N: 12     Multiple R: 1.000      Squared Multiple R: .999
   Adjusted Squared Multiple R: .999          Standard Error of Estimate: 1.976
```

Variable	Coefficient	Std Error	Std Coef	Tolerance	T	P(2 Tail)
A	1.525	0.010	1.000	.100E+01	147.898	0.000

Analysis of Variance

Source	Sum-of-Squares	DF	Mean-Square	F-Ratio	P
Regression	85440.033	1	85440.033	21873.709	0.000
Residual	42.967	11	3.906		

The prediction equation is $\hat{y} = 1.525A$ where $\beta = 1.525$ is the slope of the line. The MSE is listed in the ANOVA portion of the table as MSE = 3.906. Both of these computations can be done easily by hand if you would like to check the accuracy of the computer. Then,

$$S/N_D = 10 \log_{10} \frac{\beta^2}{MSE} = 10 \log_{10} \frac{(1.525)^2}{(3.906)}$$

$$= 10 \log_{10}(.5953)$$
$$= 10(-.2252)$$
$$= -2.252$$

The computations for the remaining runs are identical. The results for all runs are:

Run	Stop	Cup	35° R	W	S	70° R	W	S	β	MSE	S/N_D
1	1	1	53	52	57	108	109	107	1.525	3.906	−2.25
			50	52	53	108	105	105			
2	1	2	35	36	34	85	87	86	1.178	35.58	−14.09
			34	33	33	84	86	88			
3	2	1	45	55	57	95	105	107	1.431	22.06	−10.32
			46	48	52	96	98	99			
4	2	2	35	36	38	85	82	88	1.165	23.40	−12.36
			30	40	40	86	80	81			

STEP 1:   AVERAGE MARGINALS OF S/N

	STOP	CUP	STOP × CUP	
AVG 1	−8.17	−6.3	−7.3	BEST SETTINGS ARE
AVG 2	−11.30	−13.2	−12.2	STOP = 1, CUP = 1

STEP 2:   ADJUST PULL ANGLE AS REQUIRED TO REACH TARGET

9.5 Put the Process on Target and Confirm

9.5.1 Blended Approach

The prediction equation given in 9.4.1 can be readily used to hit the applicable target.

9.5.2 Taguchi Approach

Using the Taguchi approach, we find the best settings for the control factors are stop at level 1 and cup location at level 1. In Step 2 we can either adjust the pull angle as required to reach the required target (not very efficient) or we can solve the following equation:

$$\hat{y} = \beta m, \text{ or}$$
$$\text{distance (predicted)} = \beta \text{ (pull back angle)}$$
(where β is calculated from the confirmation runs)

For example, suppose our confirmation run data is (using cup = 1, stop = 1):

35°			70°			
R	W	S	R	W	S	β
52	53	54	109	108	107	1.537

If we wished to hit a target of 92 inches, then:

$$92 = 1.537 \text{ (pull back angle)}$$
$$\Rightarrow \text{pull back angle is } 59.8°$$

9.6 An Example

Let's finish this chapter with an example of experimental design on a dynamic process. In this example we will look at rolling aluminum sheets [2]. Using our process model,

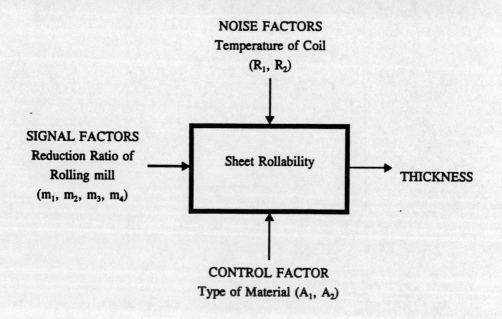

our objective is to select the best material type to obtain the least variation in thickness at all four reduction ratios.

Using a Taguchi setup for this problem, the signal and noise factors will be placed in the outer orthogonal array. The control factor will be placed in the inner array. Thickness readings for the experiment are:

Type of Material	m_1		m_2		m_3		m_4	
	R_1	R_2	R_1	R_2	R_1	R_2	R_1	R_2
A_1	102,103	110,98	86,86	88,73	58,58	64,56	34,37	40,40
A_2	85,85	86,86	77,77	78,79	69,72	70,70	65,64	63,62

Before conducting analysis of the data, let's draw a picture of the data.

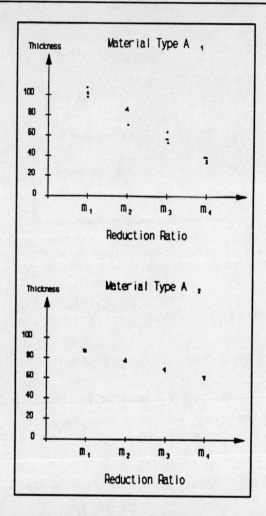

Figure 9.4 Signal Factor vs Response

From the above, it is apparent there is less variability using material Type A_2 than material Type A_1.

In the above setup, the control factor (type of material) is in the inner array and the signal and noise factors are in the outer array. For each inner array row, the applicable S/N ratio will be calculated as discussed on page 8-12, the general form of the dynamic S/N ratio is:

$$S/N_D = 10 \log_{10} \frac{\beta^2}{MSE}$$

where β is simply the least squares regression slope value b_1, (see page 4-24) and MSE is $\frac{SSE}{m-2}$ as discussed on page 4-23. Since the response value is obviously not 0 when the reduction ratio is zero, an equation of the form $\hat{y} = b_0 + b_1 x$ is fit to the data. Since Taguchi was most concerned about variability over the response range from 40 to 100, he recoded the response values by subtracting 70 from each point and then conducting regression analysis (when recoding the data for regression purposes Taguchi has no hard and fast rules for response values). In coded form, the data for the example is:

Table 9.2 Regression Input Table

A_1		A_2	
m	Result	m	Result
1	32	1	15
1	33	1	15
1	40	1	16
1	28	1	16
2	16	2	7
2	16	2	7
2	18	2	8
2	3	2	9
3	−12	3	−1
3	−12	3	2
3	−6	3	0
3	−14	3	0
4	−36	4	−5
4	−33	4	−6
4	−30	4	−7
4	−30	4	−8
$\hat{y} = 56 + -22.075(m)$		$\hat{y} = 22.625 - 7.35(m)$	
MSE = 20.737		MSE = 1.039	

For material A_1:
$\beta = -22.075$
MSE = 20.737

$$S/N_D = 10 \log_{10} \frac{\beta^2}{MSE} = 13.7$$

For material A_2:
$\beta = -7.35$
MSE = 1.039

$$S/N_D = 10 \log_{10} \frac{\beta^2}{MSE} = 17.2$$

Since S/N_D is greater for A_2 than A_1, A_2 is selected as the best. The conclusion reached was the same reached with the graphical approach.

The variability reduction problem relating to the aluminum sheet rollability example could also be addressed using ln s. Reconfiguring the data as shown in Table 9.3:

Table 9.3 The Blended Approach

Reduction Ratio	Type of Material	R_1	R_2	y	s	Ln s
M_1 (−3)	A_1 (−)	102,103	110,98	103.3	4.99	1.61
M_1 (−3)	A_2 (+)	85,85	86,86	85.5	.58	−.55
M_2 (−1)	A_1 (−)	86,86	88,73	83.3	6.89	1.93
M_2 (−1)	A_2 (+)	77,77	78,79	77.8	.96	−.04
M_3 (1)	A_1 (−)	58,58	64,56	59.0	3.46	1.24
M_3 (1)	A_2 (+)	69,72	70,70	70.3	1.26	.23
M_4 (3)	A_1 (−)	34,37	40,40	37.8	2.87	1.06
M_4 (3)	A_2 (+)	65,64	63,62	63.5	1.29	.25
Ln S −	1.46					
Ln S +	−.03					
|Δ|	1.49					

Analysis of ln s indicates we obtain reduced variability with material A_2.

Chapter 9 Bibliography

1. American Supplier Institute, Five-day Advanced Taguchi Methods Seminar. Dearborn, Michigan: February, 1992.

2. Taguchi, G., *System of Experimental Design*. Dearborn, Michigan: ASI.

CASE STUDY III-1
Robust Design
Submitted by Richard Shaw

The model XYZ Feeder Diverter Blade (located in the photocopy machine in Figure III.1) has excessive bow (see enlargement of blade bottom in Figure III.2). Using their current process set points, the vendor was unable to hold the specified part tolerance. The part in question was in the pilot manufacturing stage of an injection molding process. A histogram from initial inspection data on the part is displayed in Figure III.3.

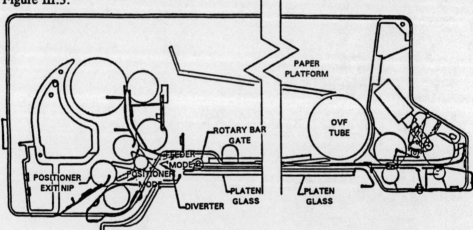

Figure III.1 Diverter Blade Location in Photocopy Machine

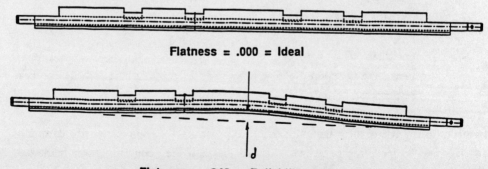

Figure III.2 Enlargement of Blade and Blade Bow

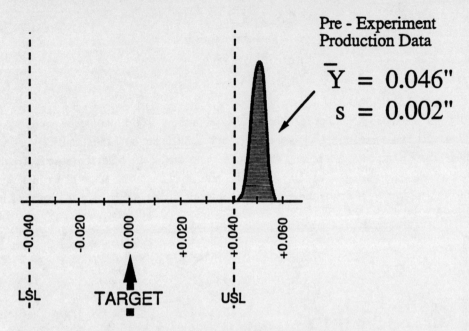

Figure III.3 Histogram of Current Blade Flatness

Several options were initially suggested as a means of addressing the flatness problem. Some of the options suggested were:

- secondary straightening
- machining
- redesign the part and use of sheet metal
- re-tool
- institute a 100% sort/scrap operation
- tweak the injection molding process so as to determine if better process settings can be found

Instead of using one of these tolerance design approaches, the team (consisting of vendor and customer engineering personnel) decided to see if they could do some parameter designing and make the process robust. As shown in Table III.1, seven control factors were identified:

Table III.1 Factors and Levels

Press Parameters	Current Settings	Possible Range	Levels		
Melt (melt) Temperature	580°F	580°F to 620°F	580	600	620
Stationary Mold (sta)	150°F	75°F to 200°F	80	140	200
Moveable Mold (Mov)	150°F	75°F to 200°F	80	140	200
Cure Time (cure)	30 sec	20 to 40 sec	20	30	40
Hold Time (hold)	6 sec	4 to 6 sec	4	5	6
Injection Time (inj)	7 sec	5 to 9 sec	5	7	9
Injection (stage 1) Pressure for First Stage	1650 psi	1350 to 1650	1350	1450	1550

The outer array noise factor was process variation. Four repetitions were made for each combination. The randomized L_{18} array with measured flatness values is given in Table III.2:

Table III.2 Completed L_{18} Experiment Worksheet

Run	Melt	Sta	Mov	Cure	Hld	Inj	Stg 1	Flatness
1	580	80	80	20	4	5	1350	52.1, 53.8, 47.1, 45.2
10	580	80	200	40	5	7	1350	45.2, 44.6, 44.2, 42.8
11	580	140	80	20	6	9	1650	39.5, 39.2, 42.3, 45.1
2	580	140	140	30	5	7	1650	53.7, 49.4, 51.0, 48.2
12	580	200	140	30	4	5	1950	38.1, 34.4, 34.9, 34.6
3	580	200	200	40	6	9	1950	42.4, 41.8, 40.3, 43.0
4	600	80	80	30	5	9	1950	32.8, 32.9, 30.6, 33.3
13	600	80	140	40	4	9	1650	36.8, 40.0, 37.3, 36.5
5	600	140	140	40	6	5	1350	46.5, 46.4, 44.0, 46.8
14	600	140	200	20	5	5	1950	42.5, 39.2, 44.5, 43.0
15	600	200	80	30	6	7	1350	39.4, 39.7, 38.2, 37.7
6	600	200	200	20	4	7	1650	38.1, 37.7, 43.5, 39.5
7	620	80	140	20	6	7	1950	47.4, 47.8, 42.1, 48.3
16	620	80	200	30	6	5	1650	60.5, 51.0, 53.3, 58.0
17	620	140	80	40	4	7	1950	48.5, 48.6, 46.8, 50.7
8	620	140	200	30	4	9	1350	40.3, 43.0, 41.4, 45.8
9	620	200	80	40	5	5	1650	41.3, 40.4, 42.8, 41.3
18	620	200	140	20	5	9	1350	32.2, 33.8, 34.9, 34.3

The data was analyzed using a Nominal is Best (Type II) S/N ratio. With Nominal is Best (Type II) the transformation equation is:

$$S/N_N \text{ (type II)} = -10 \log_{10} \frac{1}{s^2}$$

where

$$S = \sqrt{\frac{\Sigma(\bar{x}-x_i)^2}{n-1}}$$

Table III.3 contains the applicable statistics for the 18 run orthogonal array.

Table III.3 S/N Ratio and Mean per Run

Run	S/N Ratio	Mean
1	-12.1740	49.550
2	-7.525	50.575
3	-1.2791	41.875
4	-1.7221	32.400
5	-2.2423	45.925
6	-8.4592	39.700
7	-8.9671	46.250
8	-7.5657	42.625
9	0.0436	41.450
10	-0.1703	44.200
11	-8.8248	41.525
12	-4.8382	35.500
13	-4.0881	37.650
14	-8.6392	42.050
15	0.4096	38.750
16	-12.7246	55.700
17	-4.0654	48.650
18	-1.2710	33.800

Computing the level averages for each factor using the S/N ratio:

Table III.4 Level Averages for S/N_N

Factor	Highest	Lowest	Difference
Cure	-1.9669	-8.0559	6.0890
Sta	-2.5657	-6.6410	4.0753
Hold	-3.2140	-6.8651	3.6511
Stage 1	-3.8356	-6.9297	3.0941
Inj	-4.1251	-6.7625	2.6373
Mov	-4.3889	-6.4730	2.0842
Melt	-4.1236	-5.8019	1.6784

Doing the same for the mean:

Table III.5 Level Averages of the Raw Data

Factor	Highest	Lowest	Difference
Inj	45.0292	38.3125	6.7167
Sta	45.2250	38.5125	6.7125
Melt	44.7458	39.4125	5.3333
Hold	45.0042	40.7458	4.2583
Stage 1	44.4333	41.1208	3.3125
Mov	44.3583	41.6167	2.7417
Cure	43.2917	42.1458	1.1458

Figure III.3 provides a plot of the level averages for S/N_N (type II).

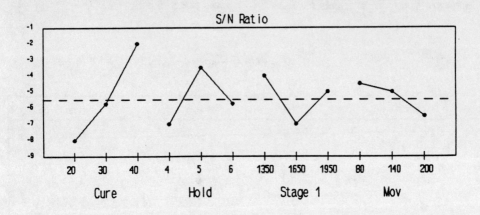

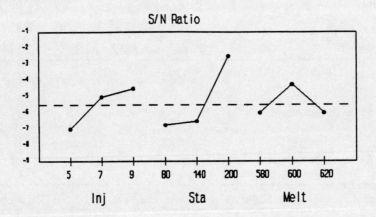

Figure III.3 Plot of Level Averages for S/N_N (Type II)

From this figure it appears that cure, sta, and hold have the largest effect on S/N_N (type II). Figure III.3 also suggests the best settings for each is:

Cure = 40 sec
Sta = 200°F
Hold = 5 sec

The other factors do not appear to be as important. Figure III.4 provides a plot of averages for each factor using the raw data.

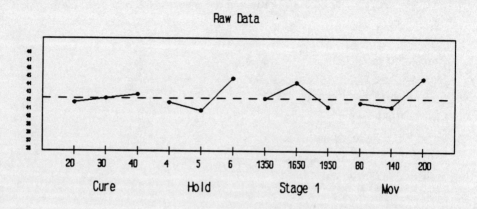

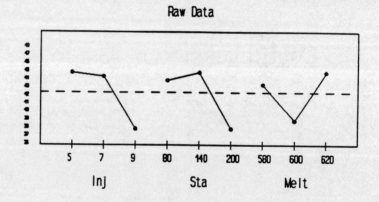

Figure III.4 Plot of Level Averages for Raw Data

This figure suggests that Inj, Sta, Melt, and Hold have the largest effect upon the average of the raw data. The best settings for these factors appear to be:

Inj = 9 sec
Sta = 200°F
Melt = 600°F
Hold = 5 sec

8 Straight Talk on Designing Experiments

Combining our analysis from the S/N ratio and the mean of the raw data, the best settings should be:

Cure = 40 sec (S/N)
Hold = 5 sec (S/N and reduced mean)
Stage 1 = 1650 psi (S/N)
Mov = 80°F (S/N)
Inj = 9 sec (reduced mean)
Sta = 200°F S/N and reduced mean)
Melt = 600°F (reduced mean)

For confirmation purposes, 30 parts were produced at the above settings. Figure III.5 displays the results:

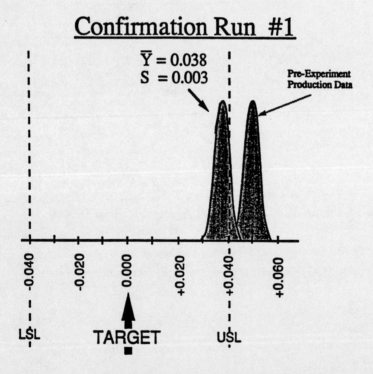

Figure III.5 Histogram of Data Collected During Confirmation Run #1

As shown in the graphic, minor improvement was obtained, but more was required. After further review of the results, the team noticed that a high stationary mold temperature appeared to have a big effect on the average and S/N. Because of this, a follow-up set of tests was conducted with stationary mold temperature increased to 225°F and all other factors set at the same levels as in the confirmation tests. An additional 30 parts were made at these settings. The results are displayed in Figure III.6 as confirmation run #2.

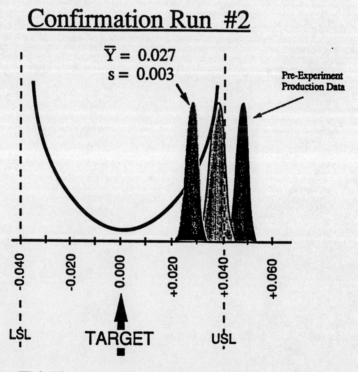

Figure III.6 Histogram of Data Collected During Confirmation Run #2

The process now appears to be able to meet the specified conditions for flatness.

IV. Modeling Designs

CHAPTER 10
CENTRAL COMPOSITE DESIGNS

Although there are a few designs that are used for modeling, we will only address Central Composite Designs (CCDs). CCDs are a very useful family of designs which are used frequently. Developed in the 1950's by Box/Wilson [1], they are frequently referred to as response surface method (RSM) designs. There are two subsets within the CCDs:

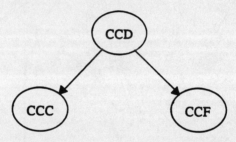

CCC is the central composite circumscribed design and CCF is the central composite faced design. The choice will depend on your experimental region and possible levels for your factors.

10.1 Planning

You must still plan your experiment as we discussed in Chapter 1. Assuming you have accomplished steps A through H listed on pages 1-1 through 1-11, you are ready to select your factors and the associated levels. CCDs can be used with two or more factors. Frequently they are conducted with five or fewer. Since you have decided to look into a CCD, you must have some reason to suspect a nonlinear relationship between your factors and responses. Since this is one motivation for choosing a modeling design, let's examine this more closely. Using our reliable statapult we conducted a simple

10-2 Straight Talk on Designing Experiments

demonstration. Using a constant pullback angle (on this particular statapult it was 4 cm on the linear measuring scale attached to the device), we fired a small rubber ball four times at each stop setting. Figure 10.1 depicts the average trajectory for each stop position:

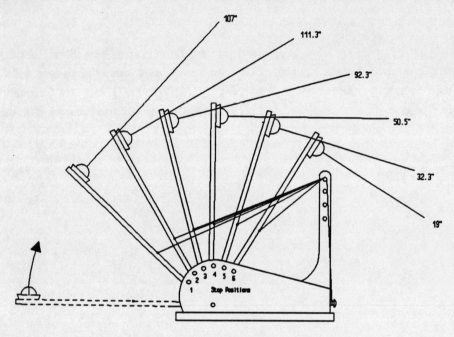

Figure 10.1 Statapult Trajectories for each Stop Position

Our data looks like:

	Stop Position					
	1	2	3	4	5	6
	108	112	92	50	34	21
	104	111	93	51	31	19
	108	110	92	50	32	18
	108	112	92	51	32	18
$\bar{y}$	107	111.25	92.25	50.5	32.25	19

We also kept the peg in position 4, the cup in position 6, and the hook in position 5. Plotting the averages, we have:

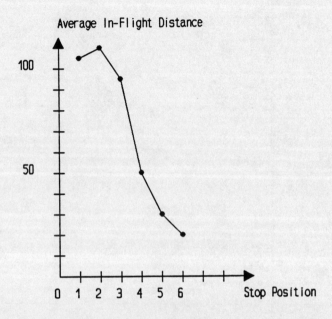

Figure 10.2 Nonlinear Relationship Between Statapult Stop Position and In-flight Distance

Clearly, the relationship between the in-flight distance and the stop position is nonlinear. If we allow the pull back angle to vary, the situation becomes more nonlinear.

Only you will know if your process exhibits nonlinearities. If so, and you wish to model it with a nonlinear equation, you must identify a low, mid-, and high level for each factor. This will be enough to allow you to use a CCF. If something in your process necessitates looking outside this range you may consider using a CCC. Because of the mathematics involved, your factors must be continuously adjustable. You must still select low, mid-, and high levels for your factors, but you must also select a level slightly below the low setting and one slightly above the high setting. We will discuss this in the next section.

10.2 Selecting an Orthogonal Array

Generally, the CCD matrix has three portions:

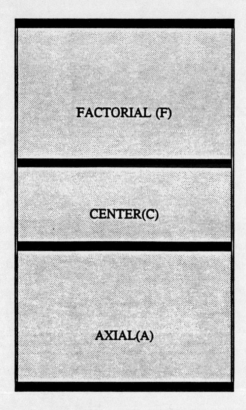

Factorial Portion: This is the familiar full or fractional factorial design at two levels.

Center Portion: This is nothing more than a predetermined number of runs with all factors set at mid-levels.

Axial Portion: These settings will be determined by the expected nonlinearity and your desire to maintain orthogonality.

For any given problem, the first two portions will remain identical. The axial portion will vary depending on your decision to use a CCF or a CCC. Using orthogonal values let's fill in the above symbolic blocks for a two-factor experiment using a four-run full factorial design for the F portion:

RUN	FACTOR 1	FACTOR 2
1	−	−
2	−	+
3	+	−
4	+	+
5	0	0
6	0	0
7	−α	0
8	α	0
9	0	−α
10	0	α

The number of runs in each portion and the information gathered is summarized in the following table:

Part	Number of Runs	Information Obtained
F	N_F (Depends upon # of factors and your strategy relative to interactions.)	• Linear effects of factors • Effects of linear interactions
C	We recommend N_C = (2 to 4) - certain software packages use more - Statgraphics suggests: $N_C = 4\sqrt{N_F} + 4 - 2 \cdot K$, where K = # of factors. See [2] for details. 2-4 centerpoints should be adequate for most applications.	• Estimate experimental error. • Estimate quadratic departure for linearity. (The centerpoints will suggest if one or more factors are non-linear.)
A	# of runs = $(N_A) = 2 \cdot K$	• Learn which factor or factors are quadratic.

10-6 Straight Talk on Designing Experiments

For the CCC design, the distance of the axial points from the center of the experimental region (in orthogonal units) is $\alpha = \sqrt[4]{N_F}$ (this will yield an orthogonal, rotatable design) [2]. For the CCF design, the axial points are ± 1 orthogonal units from the center of the experimental region.

As an example, two factor CCC and CCF designs will be set up and compared. Suppose the factors are:

Factors	Low(−)	High(+)
Time	5 minutes	9 minutes
Temp	270°F	290°F

Written in real units, the two design matrices look like:

	CCC				CCF		
Run #	Time	Temp	Portion	Run #	Time	Temp	Portion
1	5	270°F		1	5	270°F	
2	5	290°F	F	2	5	290°F	F
3	9	270°F		3	9	270°F	
4	9	290°F		4	9	290°F	
5	7	280°F	C	5	7	280°F	C
6	7	280°F		6	7	280°F	
7	4.18	280°F		7	5	280°F	
8	9.82	280°F	A	8	9	280°F	A
9	7	265.9°F		9	7	270°F	
10	7	294.1°F		10	7	290°F	

For the CCC above, $\alpha = \sqrt[4]{N_F} = \sqrt[4]{4} = 1.414$ orthogonal units.

To convert that into real units for the experimenter, we perform a simple extrapolation:

```
        -1.414   -1    0    +1  +1.414
               |----+----+----+----|         Orthogonal Scale

             ?  5    7    9  ?
  Time:       |----+----+----|                Real Scale
```

Since each orthogonal unit is equivalent to two minutes, 1.414 orthogonal units is .414(2) minutes less than 5 minutes and .414(2) minutes greater than 9 minutes. Therefore, the real axial values for time are 4.18 and 9.82 minutes, respectively.

```
             ? 270  280  290 ?
  Temp:       |----+----+----|                Real Scale
```

Since each orthogonal unit is equivalent to 10 degrees, 1.414 orthogonal units will be .414(10) degrees less than 270 degrees and .414(10) degrees more than 290 degrees. Therefore, the real axial values for temperature are 265.9 and 294.1 degrees, respectively.

For the CCF, you will recall that the axial values are the real low and high levels for each factor.

As shown in this example, the CCF provides three levels for each factor. The CCC provides five levels for each factor. The choice of conducting a CCC vs. a CCF boils down to two issues:

(1) Constraints of the experimental region you wish to explore.
(2) Statistical precision - with the choice of the proper axial point distance and number of center points - the CCC has slightly better statistical properties [1].

Now, let's see if we can set up a more complex CCC:

Example 10.1: Set up a central composite circumscribed design in an experiment where our objective is to determine the optimal time/temperature profile for a type of composite material being considered for a new missile. Our response is strength of the final product. Factors to be considered are:

10-8 Straight Talk on Designing Experiments

Factor	Low(−)	High(+)
(H) Heat up rate	2°F/min	4°F/min
(CT) Cure temperature	275°F	325°F
(D) Dwell time	90 min	150 min
(V) Vacuum	13.25" Hg	19.75" Hg
(CD) Cool down	3.25°F/min	7.75°F/min

Using a 16 run (fractional factorial) design for the "F" portion provides the following array:

Run #	H	CT	D	V	CD	
1	2	275	90	13.25	3.25	
2	2	275	90	19.75	7.75	
3	2	275	150	13.25	7.75	
4	2	275	150	19.75	3.25	
5	2	325	90	13.25	7.75	
6	2	325	90	19.75	3.25	F
7	2	325	150	13.25	3.25	
8	2	325	150	19.75	7.75	
9	4	275	90	13.25	7.75	
10	4	275	90	19.75	3.25	
11	4	275	150	13.25	3.25	
12	4	275	150	19.75	7.75	
13	4	325	90	13.25	3.25	
14	4	325	90	19.75	7.75	
15	4	325	150	13.25	7.75	
16	4	325	150	19.75	3.25	
17	3	300	120	16.50	5.50	C
18	3	300	120	16.50	5.50	
19	1	300	120	16.50	5.50	
20	5	300	120	16.50	5.50	
21	3	250	120	16.50	5.50	
22	3	350	120	16.50	5.50	
23	3	300	60	16.50	5.50	A
24	3	300	180	16.50	5.50	
25	3	300	120	10.00	5.50	
26	3	300	120	23.00	5.50	
27	3	300	120	16.50	1.0	
28	3	300	120	16.50	10.0	

Selection of the axial points is typically done by a computer. Take time now to confirm the axial points provided by the computer.

10.3 Conducting the Experiment and Analyzing Your Results

For a CCD, conducting and analyzing an experiment go hand-in-hand and illustrate one of the advantages of this RSM design — that experimentation may be done sequentially. Satisfactory results after the first two portions will allow you to terminate the experiment, saving time and money. For example, suppose we conducted an experiment on a friction welding process. Friction welding is a somewhat crude process in which two soft metals (aluminum is frequently used) are slammed together. Prior to engaging the two pieces with a given pressure, one or both parts will be turning at a given number of revolutions per minute. Suppose the response of interest is weld upset (amount the parts weld together). The results of the experiment are:

Run	Rpm	Prebond Distance	Upset	
1	3000	1	.50	
2	3000	2	.52	F
3	4000	1	.81	
4	4000	2	.83	
5	3500	1.5	.74	C
6	3500	1.5	.76	
7	3000	1.5	.51	
8	4000	1.5	.81	A
9	3500	1	.72	
10	3500	2	.76	

Initially, analysis will be conducted on the "F" portion (first 4 runs) of the array:

10-10 Straight Talk on Designing Experiments

Run	(R) Rpm	(P) Prebond Distance	R × P	Upset
1	3000	1	+	.50
2	3000	2	−	.52
3	4000	1	−	.81
4	4000	2	+	.83
Avg "Low"	.51	.655	.665	
AVG "High"	.82	.675	.665	

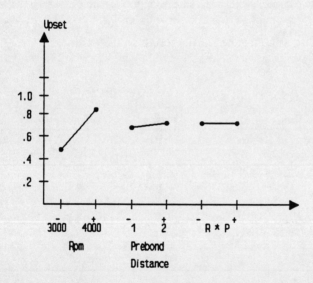

Figure 10.3 Level Averages for Friction Welding Process

Analysis of the plot of average marginals suggests Rpm may have a relatively large positive effect, prebond distance seems to have a relatively small positive effect, and there appears to be no interaction. Continuing, we add the average of the two centerpoint runs to the plot of averages.

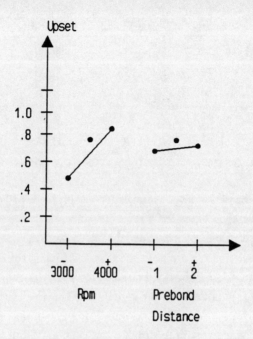

Figure 10.4 Adding Centerpoints to the Plot of Averages

If the average of the centerpoints falls on the line (or very near), it is probably safe to conclude there is no quadratic departure from linearity (software packages test for this mathematically — for a discussion of "lack of fit tests," see [3]). Since the average of the centerpoints is far from the line in our plot, it appears one or more of the factors is non-linear. We will not really know which is non-linear, however, unless we use the data from the "A" portion of the design. At this point we have gone as far as we will with simple analysis and will use regression analysis to complete the analysis. A typical output table is shown on the following page.

    The "source" column identifies the specific effect, the "coefficient" column can be thought of as the half-effect (Chapter 4), and the "probability" column provides us with a measure of importance of a particular effect. The smaller the probability value, the more likely the effect in question is important. Different people use different rules of thumb on how small the probability value must be before they conclude an effect is important. Typical values are less than 0.10 or 0.05.

Source	Coefficient	Probability
Constant	.745	0.00
R	.153	0.00
P	.013	0.04
R^2	−.081	0.00
P^2	−.001	0.85
R×P	0.00	1.00

Let's contrast the plot of level averages for this problem with the regression output table. The plot of averages suggested a relatively strong positive effect for Rpm with a relatively smaller positive effect for prebond distance. Note the coefficients for Rpm (R) and prebond distance (P). Both are positive. The coefficient for R is also much larger than the coefficient for P. Additionally, both the plot of level averages and the regression table suggest the R×P interaction is not important. The final question which must be answered is which factor (or factors) is quadratic? Review of the regression output table suggests only Rpm is important in this regard. In summary, our analysis suggests our best representation of what is occurring in the experiment is:

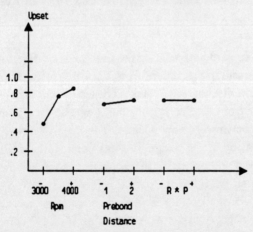

Figure 10.5 Summary Plot of Averages

From the regression output table, it appears the applicable prediction equation for upset is:

$$y = .745 + .153R + .013P - .081R^2$$

You should recall that we can draw contour maps and response surface plots using this equation. The following response surface and corresponding contour plot were drawn with RS Discover:

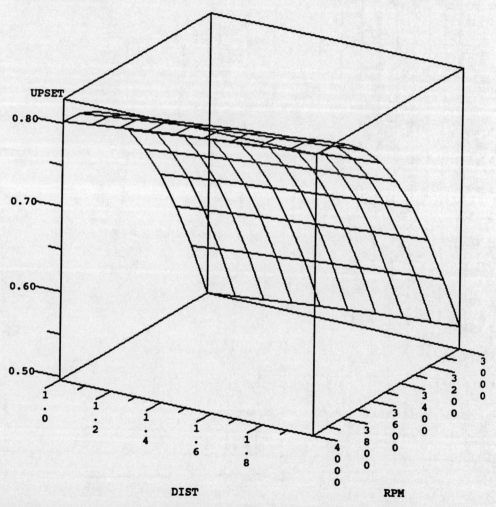

Figure 10.6 Response Surface for Friction Weld Example

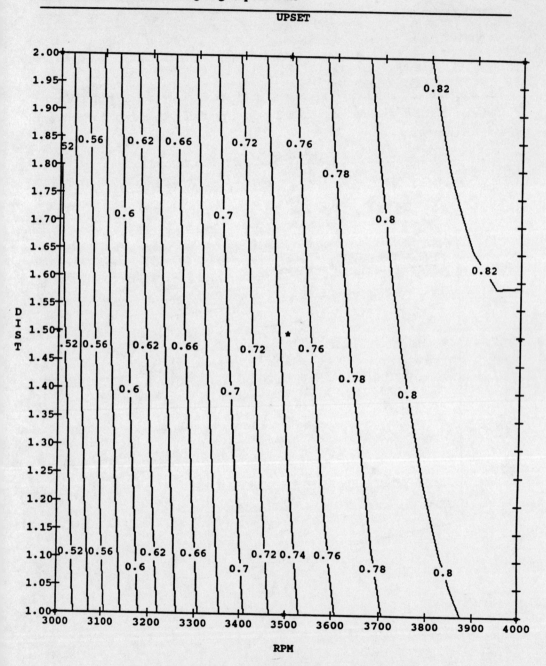

Figure 10.7 Contour Plot for Friction Weld Example

If we were interested in achieving a target weld upset of .745, we can see that an RPM of 35 and a prebond distance of 1.5 should put us right on target. The only remaining task is to CONFIRM our model.

10.4 Conclusion

Perhaps the best conclusion is to give you our summary of the advantages and disadvantages of CCDs and let you make the call.

The primary advantages are:

1. Allow for the estimation of linear & quadratic effects of factors.

2. Provide flexibility in estimation of interaction effects. All, some, or no linear interaction effects can be estimated depending upon how they are set up.

3. If desired, only the first two portions [F + C] can be run (thus saving runs). If analysis of the F + C portions indicate no quadratic departure from linearity, the last portion [A] need not be conducted. In situations where experimentation is expensive, the savings incurred may be substantial.

The primary disadvantages are:

1. In order to conduct all three parts of the design, continuously adjustable factors (quantitative) are required.

2. Without the availability of a computer program, they can be more complex to set up and analyze.

Chapter 10 Bibliography

1. Box, George E.P. and Draper, Norman R., *Empirical Model-Building and Response Surfaces*. John Wiley and Sons, Inc., 1987.

2. Myers, Raymond H., *Response Surface Methodology*. Virginia Polytechnic Institute, 1976.

3. Schmidt, S.R. and Launsby, R.G., *Understanding Industrial Designed Experiments (3rd edition)*. Colorado Springs, CO: Air Academy Press, 1991.

CASE STUDY IV-1
Central Composite Design
Submitted by Ron Boehly

To practice experimental design techniques, we decided to apply a Central Composite Design to our golf game. Five factors were considered:

Table IV.1 Factors and Levels

FACTOR	LEVELS	UNITS
(F) Feet Position	-4, -2, 0, 2, 4	Inches
(S) Back Swing	135, 180, 225, 270, 315	Degrees
(C) Club Selection	3, 4, 5, 6, 7	
(P) Placement	0, 3, 6, 9, 12	Inches
(H) Open/close Head	1, 2, 3, 4, 5	

Figure IV.1 is a diagram of the levels chosen for each factor. Two responses (distance and angle-off-center ("degrees")) were considered. Table IV.2 is a regression output table for distance. As expected, Factor C (club) had the greatest effect on distance. The coefficient is also negative, suggesting that as club head becomes larger, the distance is less. R^2 is 89.8, suggesting that approximately 90% of the variability in distance is explained by the model.

Feet Position:

BackSwing Angle:

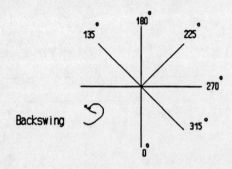

Club Selection: 3 , 4 , 5 , 6 , 7

Placement of Ball:

Club Head:

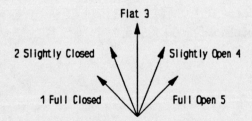

Figure IV.1 Factors and Levels

Table IV.2 Regression Table for Response = Distance

Term	Coeff	4 Signif
1	130.293	
F	−1.167	
S	4.917	
C	−10.917	
P	−2.417	
H	−0.333	
F*S	−2.625	0.1249
F*C	−3.000	0.0827
F*P	−6.750	0.0007
S*C	−3.000	0.0827
S*H	3.625	0.0399
C*H	−4.250	0.0185
F**2	2.047	0.1116
C**2	−4.203	0.0032

Contour Plots and Response Surface Plots are given in Figures IV.2 and IV.3 for important terms. The response "degrees" is evaluated in Table IV.3:

Table IV.3 Regression Table for Response = Degrees

Term	Coeff	4 Signif
1	11.952	
F	5.250	
S	−5.917	0.0001
C	−0.167	
P	3.917	
H	6.750	
F*C	4.125	0.0024
F*P	3.125	0.0151
F*H	−5.125	0.0004
C*P	2.750	0.0292
C*H	−3.250	0.0121
F**2	−1.661	0.0738
C**2	−4.661	0.0001
H**2	−4.286	0.0001

As might be expected, Factor H (club head) has the biggest effect.

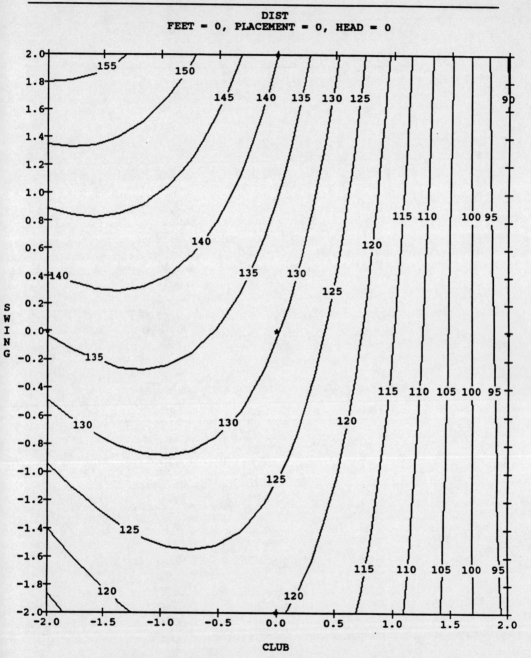

Figure IV.2 Contour Plot of Distance (F=P=H=0)

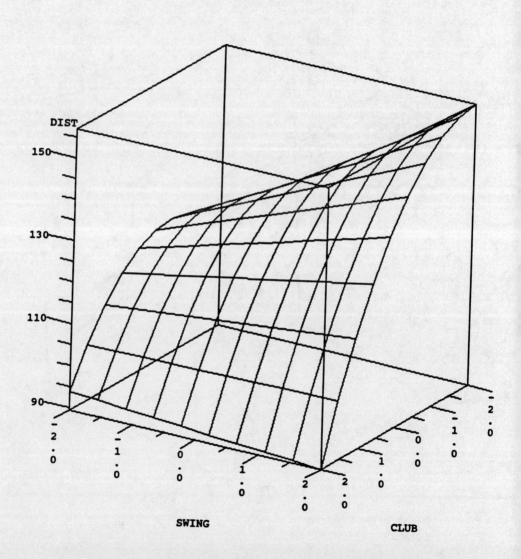

Figure IV.3 Response Surface Plot of Distance (F=P=H=0)

Figure IV.4 is the response surface for degrees using only important terms:

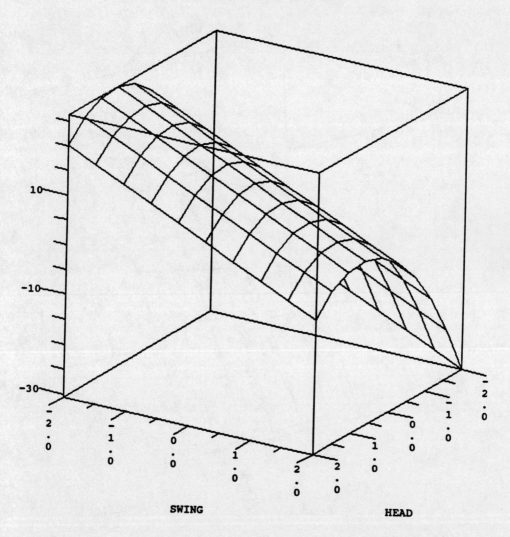

Figure IV.4 Response Surface Plot for Response=Degrees (F=C=P=0)

Combining the contour plots for both responses (Figure IV.5), we can locate feasible regions for optimizing both concurrently. Figure IV.5 is sketched after setting the three less important main factors at nominal levels (F=C=P=0).

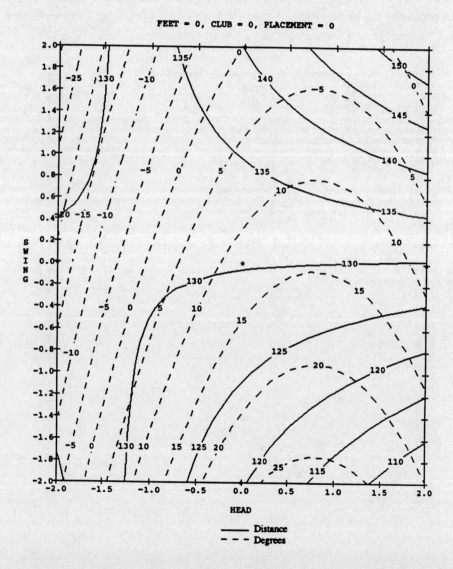

Figure IV.5 Combined Contour Plot of Distance and Degrees (F=C=P=0)

8   Straight Talk on Designing Experiments

To use Figure IV.5, let's suppose we want to hit a shot 135 yards downrange with an angle of 0 degrees off-center (no fade or hook). We must first locate the contour lines corresponding to 135 yards downrange and 0 degrees off-center. The feasible region will be the point of intersection of these two lines. The values of S and H that correspond to this point should put us on target. Figure IV.5 suggests a backswing of approximately 1.4 (290 degrees) and a head setting of −.4 (2.6) should do it (Figure IV.5.a).

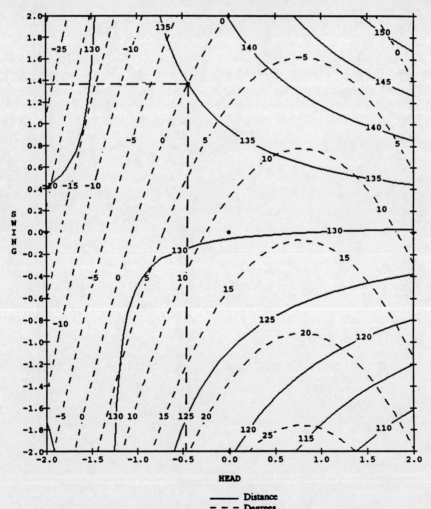

Figure IV.5.a Putting Distance and Degrees On Target

Plots such as those in Figure IV.5 are helpful in allowing us to co-optimize responses. To confirm our model, we would head out to the driving range and tee up with our feet parallel to the path of the ball, the ball directly in front of our forward foot, and a five iron in our hands. Then, we'd bring the club back about 290 degrees and hit the ball with the club head somewhere between flat and slightly closed. If the ball hits 135 yards downrange on-center, we'll pat ourselves on the back.

APPENDIX A

$$F_o = 2\left(\frac{F_R - \overline{F_R}}{d_R}\right)$$

F_o = orthogonal (coded) value for a factor (variable)
F_R = real value for a factor (variable)
$\overline{F_R}$ = average of the real ranges for a factor
d_R = difference between the real high and low values of factor

In our example on page 4-12, $A_o = 2\left(\dfrac{A_R - \overline{A_R}}{d_R}\right) = 2\left(\dfrac{A_R - 465}{-50}\right)$

$$B_o = 2\left(\frac{B_R - \overline{B_R}}{d_R}\right) = 2\left(\frac{B_R - 3}{4}\right)$$

$$C_o = 2\left(\frac{C_R - \overline{C_R}}{d_R}\right) = 2\left(\frac{C_R - 2.5}{3}\right)$$

$$A_o B_o = \left(2\left(\frac{A_R - 465}{-50}\right)\right)\left(2\left(\frac{B_R - 3}{4}\right)\right) = \frac{-4}{200}(A_R - 465)(B_R - 3)$$

$$A_o C_o = \left(2\left(\frac{A_R - 465}{50}\right)\right)\left(2\left(\frac{C_R - 2.5}{3}\right)\right) = \frac{-4}{150}(A_R - 465)(C_R - 2.5)$$

$$B_o C_o = \left(2\left(\frac{B_R - 3}{4}\right)\right)\left(2\left(\frac{C_R - 2.5}{3}\right)\right) = \frac{4}{12}(B_R - 3)(C_R - 2.5)$$

$$A_o B_o C_o = \left(2\left(\frac{A_R - 465}{50}\right)\right)\left(2\left(\frac{B_R - 3}{4}\right)\right)\left(2\left(\frac{C_R - 2.5}{3}\right)\right)$$
$$= \frac{-8}{600}(A_R - 465)(B_R - 3)(C_R - 2.5)$$

and the real $\hat{y}$ is:

$$\hat{y} = 104.521 - 32.187(2)\left(\frac{A_R - 465}{-50}\right) + 27.688(2)\left(\frac{B_R - 3}{4}\right) - 9.354\left(\frac{-4}{200}\right)$$

$$(A_R - 465)(B_R - 3) + 24.729(2)\left(\frac{C_R - 2.5}{3}\right) - 6.563\left(\frac{-4}{150}\right)(A_R - 465)$$

$$(C_R - 2.5) + 1.895\left(\frac{4}{12}\right)(B_R - 3)(C_R - 2.5) - 2.062\left(\frac{-8}{600}\right)(A_R - 465)$$

$$(B_R - 3)(C_R - 2.5)$$

$$= 104.521 + \frac{(32.187)A_R}{25} - 598.678 + \frac{(27.688)B_R}{2} - 41.532$$

$$+ \frac{9.354}{50}(A_R B_R - 3A_R - 465B_R + 1395) + 16.486 C_R - 41.215$$

$$+ .17501(A_R C_R - 2.5A_R - 465 C_R + 1162.5)$$

$$+ \frac{1.895}{3}(B_R C_R - 2.5B_R - 3C_R + 7.5)$$

$$+ \frac{2.062}{75}(A_R B_R C_R - 465 B_R C_R - 3A_R C_R + 1395 C_R - 2.5 A_R B_R)$$

$$(+ 1162.5 B_R + 7.5 A_R - 3487.5)$$

$$= 104.521 - 598.687 - 41.532 + \frac{9.354(1395)}{50} - 41.215 + (.17501)(1162.5)$$

$$+ \frac{(7.5)(1.895)}{3} - \frac{(3.487.5)(2.062)}{75}$$

$$+ A_R\left(\frac{32.187}{25} - \frac{3(9.354)}{50} - (.17501)(2.5) + \frac{2.062(7.5)}{75}\right)$$

$$+ B_R\left(\frac{27.688}{2} - \frac{465(9.354)}{50} - \frac{1.895(2.5)}{3} + \frac{2.062(1162.5)}{75}\right)$$

$$+ A_R B_R\left(\frac{9.354}{50} - \frac{2.062(2.5)}{75}\right)$$

$$+ C_R\left(16.486 - 465(.17501) - 1.895 + \frac{2.062(1395)}{75}\right)$$

$$+ A_R C_R\left(.17501 - \frac{3(2.062)}{75}\right) + B_R C_R\left(\frac{1.895}{3} - \frac{2.062(465)}{75}\right)$$

$$+ A_R B_R C_R\left(\frac{2.062}{75}\right)$$

$$= -203.624 + A_R(.495) - B_R(42.766) + A_R B_R(.1183) + C_R(-28.435)$$
$$+ A_R C_R(.09253) + B_R C_R(-12.153) + A_R B_R C_R(.02749)$$

Now, let's set pull back at 490, hook at 1 and peg at 1 to see if our equation matches experimental results.

$$A = 490, B = 1, C = 1$$

$$\hat{y} = -203.624 + .495(490) - 42.766(1) + (.1183)(490)(1) - 28.435(1)$$
$$+ .092538(490)(1) - 12.153(1)(1) + (490)(1)(1)(.02749)$$

$$\hat{y} = 72.349 \text{ (Error due to round-off.)}$$

The conversion of ŝ and ln s to real values is accomplished in the same manner.

APPENDIX B

In our example on page 4-35, we are proposing a model that includes a constant and seven variables. Our task is to determine the coefficients. Essentially we are saying that we expect to be able to predict each of our data values with

$$\hat{y} = b_0 + b_1A + b_2B + b_3AB + b_4C + b_5AC + b_6BC + b_7ABC$$

where b_0 is the constant and b_i is the coefficient of factor i. The array on page 4-24 was the first step in this process. If we add a column of ones, we can write our problem in matrix notation:

$$
\begin{array}{c}
\text{Run} \\
1 \\
2 \\
\cdot \\
\cdot \\
\cdot \\
24
\end{array}
\begin{array}{cccccccc}
\text{Const.} & A & B & AB & C & AC & BC & ABC
\end{array}
$$

$$
\begin{bmatrix}
1 & A_1 & B_1 & AB_1 & C_1 & AC_1 & BC_1 & ABC_1 \\
1 & A_2 & B_2 & AB_2 & C_2 & AC_2 & BC_2 & ABC_2 \\
\cdot & \cdot & \cdot & \cdot & \cdot & \cdot & \cdot & \cdot \\
\cdot & \cdot & \cdot & \cdot & \cdot & \cdot & \cdot & \cdot \\
\cdot & \cdot & \cdot & \cdot & \cdot & \cdot & \cdot & \cdot \\
1 & A_{24} & B_{24} & AB_{24} & C_{24} & AC_{24} & BC_{24} & ABC_{24}
\end{bmatrix}
\begin{bmatrix} b_0 \\ b_1 \\ b_2 \\ \cdot \\ \cdot \\ \cdot \\ b_7 \end{bmatrix}
=
\begin{bmatrix} y_1 \\ y_2 \\ \cdot \\ \cdot \\ \cdot \\ y_{24} \end{bmatrix}
$$

$$\overline{X} \qquad \overline{b} = \hat{y}$$

To solve this equation for $\overline{b}$, we would expect to divide both sides by $\overline{X}$, or in matrix terms multiply both sides by the inverse of $\overline{X}(\overline{X}^{-1})$. Our first problem is that we cannot find the inverse of a matrix unless it is square (number of rows = number of columns). The solution to this problem is to first multiply both sides by the transpose of $\overline{X}$ ($\overline{X}'$ or $\overline{X}^T$). As you recall, the transpose is formed by interchanging rows and columns:

$$\overline{X}'\overline{X}\overline{b} = \overline{X}'\overline{Y}$$

$$\begin{array}{c}\text{Constant}\\ A\\ B\\ AB\\ C\\ AC\\ BC\\ ABC\end{array}\begin{bmatrix} 1 & 1 & \cdots & 1 \\ A_1 & A_2 & \cdots & A_{24} \\ B_1 & B_2 & \cdots & B_{24} \\ \cdot & \cdot & \cdots & \cdot \\ \cdot & \cdot & \cdots & \cdot \\ \cdot & \cdot & \cdots & \cdot \\ \cdot & \cdot & \cdots & \cdot \\ ABC_1 & ABC_2 & \cdots & ABC_{24} \end{bmatrix} \begin{bmatrix} 1 & A_1 & B_1 & \cdots & ABC_1 \\ 1 & A_2 & B_2 & \cdots & \cdot \\ \cdot & \cdot & \cdot & \cdots & \cdot \\ \cdot & \cdot & \cdot & \cdots & \cdot \\ \cdot & \cdot & \cdot & \cdots & \cdot \\ 1 & A_{24} & B_{24} & \cdots & ABC_{24} \end{bmatrix} \begin{bmatrix} b_0 \\ b_1 \\ \cdot \\ \cdot \\ \cdot \\ b_7 \end{bmatrix}$$

$$= \begin{bmatrix} 1 & 1 & \cdots & 1 \\ A_1 & A_2 & \cdots & A_{24} \\ \cdot & \cdot & \cdots & \cdot \\ \cdot & \cdot & \cdots & \cdot \\ \cdot & \cdot & \cdots & \cdot \\ ABC_1 & \cdot & \cdots & ABC_{24} \end{bmatrix} \begin{bmatrix} y_1 \\ y_2 \\ \cdot \\ \cdot \\ \cdot \\ y_{24} \end{bmatrix}$$

After multiplying matrices:

$$\begin{bmatrix} \sum 1 & \sum A_i & \sum B_i & \cdots & \sum ABC_i \\ \sum A_i & \sum A_i^2 & \sum A_i B_i & \cdots & \cdot \\ \sum B_i & \sum B_i A_i & \sum B_i^2 & \cdots & \cdot \\ \sum AB_i & \cdot & \cdot & \cdots & \cdot \\ \cdot & \cdot & \cdot & \cdots & \cdot \\ \cdot & \cdot & \cdot & \cdots & \cdot \\ \cdot & \cdot & \cdot & \cdots & \cdot \\ \sum ABC_i & \cdot & \cdot & \cdots & \sum (ABC_i)^2 \end{bmatrix} \begin{bmatrix} b_0 \\ b_1 \\ \cdot \\ \cdot \\ \cdot \\ b_7 \end{bmatrix} = \begin{bmatrix} \sum y_i \\ \sum A_i y_i \\ \cdot \\ \cdot \\ \cdot \\ \sum (ABC)_i y_i \end{bmatrix}$$

$$\overline{X}'\overline{X} \quad \cdot \quad \overline{b} \quad = \quad \overline{X}'\overline{Y}$$

Matrix Solution to Least Squares Method — Appendix B

Since our design matrix was orthogonal, $\Sigma A_i = \Sigma B_i = \Sigma AB_i = \ldots = \Sigma(ABC)_i = 0$ and the only nonzero terms are those on the diagonal. A diagonal $\bar{X}'\bar{X}$ matrix defines an orthogonal design.

$$\begin{bmatrix} n & & & & & & & \\ & n & & & & & & \\ & & n & & & 0 & & \\ & & & \cdot & & & & \\ & & & & \cdot & & & \\ & & & & & \cdot & & \\ & & 0 & & & & \cdot & \\ & & & & & & & n \end{bmatrix} \begin{bmatrix} b_0 \\ b_1 \\ \cdot \\ \cdot \\ \cdot \\ \cdot \\ b_7 \end{bmatrix} = \begin{bmatrix} \sum y_i \\ \sum A_1 y_i \\ \cdot \\ \cdot \\ \cdot \\ \sum(ABC)_i y_i \end{bmatrix}$$

In this case, $n = 24$.

Once again, if we multiply matrices

$$\begin{bmatrix} nb_0 & & & & & & & \\ & nb_1 & & & & & & \\ & & nb_2 & & & & & \\ & & & \cdot & & 0 & & \\ & & & & \cdot & & & \\ & & & 0 & & \cdot & & \\ & & & & & & \cdot & \\ & & & & & & & nb_7 \end{bmatrix} = \begin{bmatrix} \sum y_i \\ \sum A_i y_i \\ \cdot \\ \cdot \\ \cdot \\ \sum(ABC)_i y_i \end{bmatrix}$$

and equate rows

$$nb_0 = \sum y_i \qquad\qquad b_0 = \frac{\sum y_i}{n}$$

$$nb_1 = \sum A_i y_i \qquad\qquad b_1 = \frac{\sum A_i y_i}{n}$$

$$\vdots \qquad\Rightarrow\qquad \vdots$$

$$nb_7 = \sum (ABC)_i y_i \qquad\qquad b_7 = \frac{\sum (ABC)_i y_i}{n}$$

Let's do a quick check on b_0 and b_1.

$$b_0 = \frac{\sum y_i}{n} = \bar{y} = 104.521 \quad\checkmark$$

$$b_1 = \frac{\sum A_i y_i}{n} = \frac{A_1 y_1 + A_2 y_2 + \ldots + A_{24} y_{24}}{24}$$

$$= \frac{\begin{array}{l}(-1)(85) + (-1)(128) + (-1)(150) + (-1)(215) + (1)(32.5) + (1)(69) \\ + (1)(68) + (1)(114) + (-1)(74) + (-1)(132) + (-1)(145) + (-1)(208) \\ + (1)(36) + (1)(80) + (1)(81.5) + (1)(101.5) + (-1)(58) + (-1)(121) \\ + (-1)(120.5) + (-1)(204) + (1)(38.5) + (1)(68) + (1)(68.5) + (1)(110.5)\end{array}}{24}$$

$$= -772.5/24 \quad = -32.1875 \quad \approx -32.188 \quad\checkmark$$

Now, let's see how these coefficients vary.

$$\sigma^2(b_0) = \sigma^2\left(\sum y_i/n\right)$$
$$= \sigma^2\left[\frac{y_1 + y_2 + \ldots + y_{24}}{n}\right]$$
$$= \sigma^2\left[\frac{y_1}{n} + \frac{y_2}{n} + \ldots + \frac{y_{24}}{n}\right]$$
$$= \sigma^2\left[\frac{y_1}{n}\right] + \sigma^2\left[\frac{y_2}{n}\right] + \ldots + \sigma^2\left[\frac{y_{24}}{n}\right]$$
$$= \frac{1}{n^2}\sigma^2(y_1) + \frac{1}{n^2}\sigma^2(y_2) + \ldots + \frac{1}{n^2}\sigma^2(y_{24})$$
$$= \frac{1}{n^2}\left[\sigma^2(y) + \ldots + \sigma^2(y_{24})\right]$$
$$\approx \frac{1}{n^2}\left[n\sigma^2(y)\right] = \frac{\sigma^2(y)}{n}$$

[Assuming the variance of y_i is the same as $\sigma^2(y)$.]

You know that our best estimate of the variance of the population is MSE. So, let $\sigma^2(y) \approx$ MSE.

$$\Rightarrow \quad \sigma^2(b_0) = \frac{\text{MSE}}{n} = \frac{81.146}{(24)} = 3.381$$

$$\Rightarrow \quad \sigma(b_0) = \sqrt{3.381} = 1.8388 \approx 1.839$$

For b_1,

$$\sigma^2(b_1) = \sigma^2\left(\sum A_i y_i / n\right)$$

$$= \sigma^2\left[\frac{A_1 y_1 + A_2 y_2 + \ldots + A_{24} y_{24}}{n}\right]$$

$$= \frac{A_1^2}{n^2}\sigma^2(y_1) + \frac{A_2^2}{n^2}\sigma^2(y_2) + \ldots + \frac{A_{24}^2}{n^2}\sigma^2(y_{24})$$

$$= \frac{n}{n^2}\sigma^2(y_1) + \frac{n}{n^2}\sigma^2(y_2) + \ldots + \frac{n}{n^2}\sigma^2(y_{24})$$

$$= \frac{1}{n^2}\left[\sigma^2(y_1) + \sigma^2(y_2) + \ldots + \sigma^2(y_{24})\right]$$

$$= \frac{1}{n^2}\left(n\sigma^2(y)\right)$$

$$= \frac{\sigma^2(y)}{n}$$

And as above, $\sigma^2(b_1) = \frac{\sigma^2(y)}{n} \approx \frac{MSE}{n}$ which in this case will be identical to b_0. The estimated standard deviations of the remaining coefficients follow identically. These are noted on the regression table as **STD ERROR** (of the coefficients).

INDEX

A

Adjusted squared multiple R, 4-40
Aliasing, 2-6
Analysis of means (ANOM), 4-1, 4-3
Analysis of variance (ANOVA), 4-1
Axial points, 10-4..10-7

B

Box and bubble chart 13, II-1, (applied to robust design) 8-1
Box-Behnken design, 2-17..2-18, 2-28
Box-Wilson design, 2-27

C

Cause-and-effect diagram, 1-7
Central composite designs, 2-17..2-18, 10-1..10-15:
    Factorial portion, 10-4..10-7
    Center portion, 10-4..10-7
    Axial portion, 10-4..10-7
Classical design matrix, 2-22
Compound noise factors, 8-6
Concept design, 7-6..7-7
Confirmation, 6-1
Confounding, 2-6
Contour plot, 5-2, 5-5
Control factors, 1-6, 7-2, 9-2

D

D-optimal design, 2-17..2-23, 2-26..2-27
Degrees of freedom (df), 2-20, 4-23
Design matrix, 2-1

Designed experiment, 5
Designs: Box-Behnken, 2-17..2-18, 2-28
  Box-Wilson, 2-27
  Central composite, 2-17..2-18, 10-1..10-15
  D-optimal, 2-17..2-23, 2-26..2-27
  Foldover, 2-17, 2-26
  Fractional factorial, 2-26
  Full factorial, 2-26..2-27
  Hadamard matrices, 2-17
  Latin Squares, 2-8, 2-17, 2-28
  Plackett-Burman, 2-17..2-18, 2-26
Dispersion, 7-3
Distributions: Z, 4-28..4-29
  t, 4-28, 4-35..4-36
  F, 4-28, 4-30..4-34
  Normal, 4-28
Dynamic signal-to-noise ratio, 8-11, 9-6..9-13

E

External customers, 1-2
External noise factors 7-3

F

F ratio, 4-30, 4-32, 4-35
F statistic, 4-34
Factors: Control, 1-6
  Noise, 5, 1-6
Foldover designs, 2-17, 2-26
Form:Steps in Conducting Experiments, II-2..II-4

Fractional factorial, 2-6, 2-18, 10-4, 10-8, 2-26
Full factorial 2-4, 2-18, 2-26, 2-27
Function, 1-2
Functional analysis, 1-2..1-4

G
Graphical analysis, 4-3
Guidelines for selecting responses, 11, 1-5

H
Hadamard matrices, 2-17

I
Inner array, 7-7..7-9, 8-3..8-4
Interaction plot, 4-9..4-11
Interactions: 2-factor, 2-5..2-7
  3-factor, 2-5..2-6
Insensitive products and processes, 6
Internal customers, 1-1

L
Lack of fit, 10-11
Latin squares, 2-17, 2-28
Linear response 1-7..1-10

M
Manufacturing/troubleshooting, 8
Main effects, 2-7
Mean square between (MSB), 4-23..4-24, 4-26..4-27
Mean square estimate (MSE), 4-20, 4-23
Mean square regression (MSR), 4-24
Modeling, 10, 10-1
Multiple R, 4-37

Multiple regression, 4-1

N
Noise factors, 5, 7-1:
  Compound, 8-6
  Degradation, 7-3..7-4
  External, 7-3..7-4
  Internal, 7-3..7-4
Non-linear response 1-7..1-9, 10-1..10-3

O
Objectives of experimental design, 8, 1-4
One-factor-at-a-time approach, 2-24..2-25
Orthogonal array, 2-22..2-24
Orthogonal designs, 2-3..2-6
Outer array, 7-7..7-9, 8-3..8-4
Over-the-wall engineering approach, 7-4

P
P value, 4-34
Parameter design, 7-6..7-7
Pareto chart, 2-10..2-11, 4-12..4-13
Partial aliasing, 2-6
Pizza Bin of Proven Technologies, 7-5
Plackett-Burman designs, 2-17..2-18, 2-26
Plot of averages, 4-7..4-9
Population standard deviation, 4-20..4-21
Population variance, 4-16..4-21
Prediction equation, 4-12

Q
QFD, 6
Q-Edge Software, 3-2

R

Randomization, 3-3
Regression output, 4-16, 4-41..4-42
Repetition, 2-21, 3-3..3-4
Replication, 2-21, 3-3..3-4
Resolution, 2-7
Response: Categorical, 1-5..1-6
  Continuous, 1-5..1-6
  Guidelines for selecting, 11, 1-5
  Linear, 1-7..1-10
  Non-linear, 1-7..1-9
Response surface, 4-21
Response surface methodology, 5-2
RS Discover Software, 3-3, 2-21
Rules of thumb, 11, 1-5, 4-14, 4-32

S

Seven realities, 4-20
Screening, 9
Signal function, 9-1..9-2
Signal to noise (S/N) ratio, 8-10..8-12
Squared multiple R, 4-37
Standard coefficient, 4-37
Standard error of the estimate, 4-20..4-21
Standard error of the coefficients, 4-35..4-37
Statapult adjustments, 1-1
Sum of squares, 4-24..4-27
Sum of squares residual (SSE), 4-20..4-23

T

T statistic, 4-34
Taguchi Tabled designs, 2-7..2-16
Three-level designs, 2-27

Tolerance designs, 4-41, 7-6..7-7, 7-10,
Two-level designs, 2-26, 4-2, 4-22
Type I error (α), 4-33..4-34
Type II error (β), 4-33